Motor Cycle Tuning (Four-Stroke)

Motor Cycle Tuning (Four-Stroke)

Second Edition

John Robinson

Newnes
An imprint of Butterworth-Heinemann Ltd
Linacre House, Jordan Hill, Oxford OX2 8DP

☙ A member of the Reed Elsevier group

OXFORD   LONDON   GUILDFORD   BOSTON
MUNICH   NEW DELHI   SINGAPORE   SYDNEY
TOKYO   TORONTO   WELLINGTON

First published 1986
Reprinted 1988, 1990, 1991
Second edition 1994

© Butterworth-Heinemann 1986, 1994

All rights reserved. No part of this publication may be reproduced in any material form (including photocopying or storing in any medium by electronic means and whether or not transiently or incidentally to some other use of this publication) without the written permission of the copyright holder except in accordance with the provisions of the Copyright, Designs and Patents Act 1988 or under the terms of a licence issued by the Copyright Licensing Agency Ltd, 90 Tottenham Court Road, London, England W1P 9HE. Applications for the copyright holder's written permission to reproduce any part of this publication should be addressed to the publishers.

ISBN 0 7506 1805 1

Printed and bound in Great Britain by
Redwood Books, Trowbridge, Wiltshire

Preface

Engine development is usually slow and expensive – although this does not mean that it is impossible. Indeed, the results can be very rewarding, both in competition and in the sheer pleasure of using a motor which is crisp and perfectly set up.

The returns are well worth the effort, but do not fall into the trap of expecting the wrong things. The manufacturers put a lot of work into development and they are very good at it. Consequently there is no reason to suppose that it will be possible to make any all-round engine improvements, other than those gained by careful assembly to the exact, stock tolerances. However, it is possible to make improvements in one area, at the expense of losses in other areas. Increases in high speed power are nearly always accompanied by losses in low-speed power and engine flexibility. Extra power is soon followed by reduced reliability, plus an increase in wear and noise. In many cases this will be an acceptable trade but it will make the machine more difficult to ride and it assumes a level of rider skill as well as a high degree of mechanical ability on the part of the engine builder.

This book does not set out to be an instruction manual; it attempts to explain how various stages of engine tune are reached and to describe typical development work, explaining some of the theory so that a feasible – practical – development programme can be devized.

The idea was to cut a pragmatic path between theory and what actually happens in real life; the phenomena described are all known to work . . . the trick is in making them all work together.

I am grateful to the many people who helped provide the material for this book, whether it was information, illustrations or simply experience.

In particular, my thanks are due to: AE (Hepolite); Amal Carburetters; Martyn Barnwell; John Carpenter (Mistral Engineering); Dell'Orto Carburettors; Patrick Gosling; Heron Suzuki GB; Honda UK; Kim Hull; Paul Ivey (Race Engine Components); Kawasaki Motors UK; Leon Moss (LEDAR); Mikuni Carburettors; NGK; Performance Bikes; Rod Sloane; Rex White; Jim Wells; Yamaha Motor Co. Ltd.

J.R.

Contents

Chapter 1:   *Basic principles*   *1*
Chapter 2:   *Tools and equipment*   *16*
Chapter 3:   *Development*   *27*
Chapter 4:   *Pistons, liners, combustion chambers*   *38*
Chapter 5:   *Valves and cams*   *51*
Chapter 6:   *Exhaust system*   *73*
Chapter 7:   *Fuel system*   *82*
Chapter 8:   *Ignition*   *105*
Chapter 9:   *Lubrication and cooling*   *114*
Chapter 10: *Engine preparation*   *121*
Chapter 11: *Fine tuning and testing*   *137*
Appendix: *Valve lift, velocity and acceleration*   *143*
             Valve time-area   *144*
             Road loads and gearing   *150*
             Compression ratio   *161*
             Piston travel v. crank rotation   *162*
             Piston velocity and acceleration   *164*
             Vibration and balance   *165*
             Cam timing v. cylinder head height   *167*
             Valve guide clearance   *167*
             Torque wrench settings   *168*
             Strength of materials   *168*
Index   *175*

Chapter 1

Basic principles

Engines make power by extracting heat from air flow. In order to do this, they must first pump the air in and out, add fuel and burn it, convert the heat into useful work and finally remove the waste products. Consequently, the engine has two basic functions, one as an air pump and the other as a heat pump or heat exchanger.

In a four-stroke piston engine the various phases are quite clearly defined and 'tuning' can have several meanings:

(1) to improve the efficiency of a particular process
(2) to alter the parts which control the process in order to increase its capacity
(3) to make two or more processes work in harmony (to make one part 'match' another; ultimately to benefit from the huge increases in efficiency when one part works in resonance with another).

The various processes which make up the four-stroke cycle are usually dependent upon speed or time (the higher the speed, the less time there is for a given event). In some cases, tuning is a battle against time: for example, to persuade more air to enter the engine at a particular speed, or to maintain the same air flow at a higher engine speed. Both mean that a certain amount of air has to be moved in a shorter period of time. The speed/time factor does not always work against us, though. As the speed is raised there is less time for losses to occur, such as heat from the combustion process being lost to the cooling system, or worse, being wasted in raising the temperature of the engine components.

Heat and temperature are not the same thing; heat is actually the same as power and is defined as a rate of doing work (or moving a force, such as lifting a weight). If heat flows into an object and has nothing else to do, then it will raise the temperature of the object. This is usually undesirable as far as engines are concerned because it also makes most materials expand, upsetting their working clearances, and the strength and fatigue life of most materials is reduced at higher temperatures (eventually metals melt and have no tensile strength at all).

In an engine the heat is used to raise the pressure of the gas; if the heat did nothing else, then the *thermal efficiency* would be 100 per cent. Unfortunately it also raises the gas temperature, and heat flows into the metal of the engine, into the coolant and into the exhaust. In practice, spark-ignition engines have a thermal efficiency of 25 to 30 per cent; in other words, only 30 per cent of the heat liberated by burning the fuel appears as useful power at the crankshaft.

This is quite easily calculated; the heat content of gasoline is quoted by the oil companies (around 20,000 BTU/lb) and the fuel flow and engine power can be measured. Typically, an engine giving 50 bhp will flow fuel at a rate of 25 lb/h which is a potential heat flow of 500,000 BTU/h and, as 1 BTU/h is equivalent to 0.000393 bhp, this fuel flow should produce 196.5 bhp. The thermal efficiency in this case is 50 (what you get) divided by 196.5 (what you have to pay for) – or 25.4 per cent.

Retrieving some of the lost 70–75 per cent would clearly be a worthwhile exercise but the result, as it stands, is that some heat does the required job of raising gas pressure, to a level P. The pressure acts on the piston(s) whose area amounts to A, so a force of PA is generated. This force pushes the piston along the bore, turning the crankshaft. During combustion the pressure level increases; as the piston moves along the bore the volume inside the cylinder increases and the pressure decreases. At no time in the cycle is P constant. The force PA is transmitted to the crankshaft via the connecting rod which, as the engine turns, makes a varying angle with the crank. It will exert most force when the angle between rod and crank is 90 degrees and this exact position depends upon the relationship between the length of the connecting rod and the length of the crank throw. The crank position at which P reaches a maximum and at which the leverage of the rod reaches a maximum should coincide in order to produce the greatest possible turning force at the crankshaft. To take it to extremes, if P reached a maximum at, or just before, TDC, then the force in the connecting rod would not try to turn the crankshaft in the required direction. The increasing pressure would then have nothing to do except raise the gas temperature further (when the ignition timing is advanced too far this is exactly what does happen).

The torque at the crankshaft depends first on the force generated at the piston (PA) and second on the angle between rod and crank, reaching a maximum at 90 degrees, when the torque will be PAS/2, where S/2 is half the stroke. As both P and the crank angle vary continuously, it helps to think of the torque delivery in terms of a steady *mean* pressure and a constant mean value for the stroke/rod factor.

Torque (T) will therefore be a function of P, A and S. If the engine is running at a speed N, its power (HP) will be a function of T and N (the definition of 1 horsepower is 33,000 ft-lbf/min) and at a speed of N rev/min, the mean torque will move through a distance $2\pi N$ every minute.

Consequently HP = $2\pi NT/33,000$
where T = f(PAS)

In order to increase torque, it will be necessary to increase A and/or S (both of which will raise the engine displacement) or to increase P.

In order to increase power, it will be necessary to increase T or N or both. Obviously, if T is reduced and N is increased by a proportional amount, the power will remain the same. However, if T is reduced but N is increased excessively, power will also increase.

Apart from increasing the engine size, the only means of increasing torque is to increase the pressure level P (and to reduce losses, which is a separate subject). This can be achieved by increasing the thermal efficiency, as mentioned above. Alternatively the pressure will be raised if more gas is introduced into the engine. Similarly, power can be raised if the same torque (i.e. air flow) can be maintained at higher crankshaft speeds. For the same engine size, both of these steps involve making some kind of improvement to the air flow, and the easiest way to look at the various possibilities is to follow the path of the air through the engine.

Intake

If a piston which displaces 250 cc is moved slowly from TDC to BDC, it will pull 250 cc of air through the intake valve; its *volumetric* efficiency would be 100 per cent. If the throttle were closed or if there were some other major obstruction, the piston would not draw in as much air; once inside, of course, it would expand to fill the 250 cc cylinder, but it would do so at reduced pressure and consequently the power produced from it would be less. Volumetric efficiency, despite its name, can only be evaluated from the mass of air flowing into an engine. If our 250 cc of air weighs x grams and the piston pulls in 0.9x grams, its efficiency will be 90 per cent. Because air is compressible, it is perfectly feasible for the piston to pull in 1.05x grams, giving an efficiency of 105 per cent. This would be trapped in the 250 cc volume, and therefore its pressure would be higher (than atmospheric) and the combustion pressure (P) would also be greater than before.

It takes time to accelerate the gas waiting in the intake as well as for the valve to reach full lift – below this level, it acts as an obstruction just like a partly-closed throttle. It also takes time to close the valve at the end of the intake stroke. This starting and stopping obviously cannot be done at full air flow, so to get full value from the intake stroke, the valve has to be opened earlier and closed later. While this may help the intake process, it also disturbs the following compression stroke and the preceding exhaust stroke.

Now, although the volumetric efficiency could easily be 100% at extremely low speeds (like 1 rev/min) it drops dramatically for a typical engine in the region of 1,000 to 4,000 rev/min. This is because the time interval is not long enough for the gas to start flowing, fill the cylinder and be stopped. Also there is a loss of efficiency because the parts are engineered to work at higher speeds and there is a total mismatch of valve size/timing at low engine speeds.

Between 4,000 and 7,000 rev/min there is an indecisive period, the volumetric efficiency depending on the engine's specification but from here to peak torque (which coincides with peak air flow and could be anywhere between 7,500 and 10,000 rev/min) the volumetric efficiency *increases*. It may even reach 120 per cent within a fairly narrow speed range at peak torque.

The time interval gets progressively smaller as the speed increases, so speed alone is not a help to air flow. One reason for the increase in efficiency is the same reason that the air flow was so bad below 4,000 rev/min; the cams are timed to open and close the valves to suit the engine speed where peak torque is required. At lower speeds they will be opening too early – there will be enough time for the new gas to mix with the exhaust gas, some new gas will be lost, some exhaust gas will remain in the cylinder. At lower engine speeds the valve will also close too late – there will be time for the incoming gas to stop and then flow back past the valve; in extreme cases, this will cause spitting back through the carburettor.

At high speeds, where the cams are designed to work, due to the shortened time interval the exhaust gas will be unable to dilute the new gas and the new gas will be unable to escape, the closing valve trapping it inside the cylinder. Above peak torque, the air flow drops again, mainly because the time interval is too short to accept the full process.

At high speeds there is another factor to be considered. The piston will also be travelling at a higher speed and will induce a greater gas velocity in the intake port. In gases, the various energy levels can be swapped around fairly easily – when the kinetic energy is increased by raising the speed, the pressure energy drops in proportion. When the gas then gets into the cylinder and stops, its velocity drops to zero and its pressure energy rises equally. Eventually this pressure rise will overcome the flow of gas into the cylinder and, ideally, the valve should close just before this happens. The trouble is that the ideal closing point is closely related to engine speed, what is right at one speed is too late for lower speeds and too early for higher speeds. Also the valve takes a certain amount of time to close, it is not instantaneous and the theoretical ideal has to be compromised.

Nevertheless, valve opening and closing are still pretty violent actions – violent enough to cause yet another phenomenon which can either help or hinder gas flow. When the valve opens suddenly, there is a drop in pressure which is transmitted as a low-pressure wave back along the incoming gas. When the valve shuts, the sudden throttling action creates high pressure, which can also be transmitted along the intake passage. When pressure waves reach the end of a pipe or a sudden change in section, they are reflected back from whence they came. A closed pipe simply reflects the same pressure wave; an open pipe reverses the reflected wave, turning high pressure into low pressure and vice versa.

If a reflected wave appears at the valve just as it opens or closes it could help gas flow; high pressure would help to initiate flow into the cylinder and would, on closing, minimize any loss past the valve. On the other hand, the appearance of a low-pressure wave would have adverse effects.

The arrival of any waves depends upon the length of the tract to the first major section change and the speed of wave propagation in the gas, both of which are substantially constant. The time interval between opening and closing, however, varies with engine speed, so any effects – beneficial or otherwise – will only occur within a limited speed range. Pressure waves can be detected by fitting a pressure-sensitive transducer into the tract and displaying its output on an oscilloscope (see Fig. 23).

The size and shape of the intake passage can obviously have a major effect on air flow, and, for similar reasons, so can the air box or air filter housing. This acts as a reservoir and can easily boost the air flow at certain speed ranges. If the air flow is increased substantially, the air box or filter could reach its capacity, above which point it simply becomes restrictive. Many roadsters have an intake silencer, a webbed trumpet or duct at the entry to the air box. This is often the most restrictive part of the intake.

Smoothly-radiussed bellmouths improve the flow into pipes, chambers, carburettors, etc, while sudden steps or changes in section can either interrupt pressure waves or set up unwanted waves. Production machines often have such devices, or surge tanks in the intake tract, in order to prevent resonant effects from making the engine too peaky.

Compression

The more the gas is compressed the easier it is to burn and the faster it burns – which is important at high engine speeds when the diminishing time interval becomes the limiting factor. More efficient heat liberation is also associated with higher compression pressures; it means that less heat is lost to the surroundings and more is concentrated in raising the gas pressure.

Theoretically, the thermal efficiency is:

$$E = 1 - (1/r)^{n-1}$$

where E = thermal efficiency
 r = compression ratio
 n = ratio of specific heats of gas at constant pressure and constant volume (ideal 1.4; less in real engines)

So if our earlier example had a thermal efficiency of 25 per cent with a compression ratio of 8:1, then raising this to 11:1 would give a thermal efficiency of 28 to 29 per cent, allowing for losses. This would give 56 bhp instead of 50 bhp. One problem associated with raising compression ratios is that because the fuel burns more easily, it will eventually ignite itself and the engine simply is not strong enough to withstand the force of the detonation (or knock) for very long.

Some cylinder head designs permit more efficient combustion and there is some evidence that swirl (or turbulence) in the gas also improves combustion. In these cases it is possible to run higher compression ratios.

Expansion

Having rapid and complete combustion is the first part, the second is to make the most efficient use of the pressure build-up in the gas. Piston and head designs which minimize surface area (and therefore heat loss) can help, as can the design of the piston skirt, rings and oil control, in optimizing the heat loss to the cylinder wall. Frictional drag as well as the angle made between rod and crank will depend on the length of the connecting rod. Shorter rods produce more side thrust but also give maximum torque closer to TDC, where the gas pressure is likely to be higher.

For a given engine running in its power band, expansion ends when the cylinder pressure has dropped to a level where its ability to move the piston is less than its usefulness in rapidly expanding into the exhaust system. At this point, the exhaust valve should open.

Exhaust

The basic need is to scavenge the cylinder completely, in time to make room for the fresh intake charge. If the valve opens early then some of the working stroke will be lost but this is offset by the advantage in starting the exhaust stroke when there is still enough gas pressure to promote quick flow. The optimum exhaust-open point is critical to the performance of the engine and it is closely related to engine speed. Rapid valve-opening plus the pressure inside the cylinder can create a strong pressure pulse which is valuable when used with a suitable exhaust system. As with the intake, pressure pulses are reflected back to the valve, an open pipe (or an increase in section) changing high pressure to low pressure and vice versa.

The exhaust system has to be proportioned so that there is low pressure in the exhaust port when the valve opens and high pressure when it closes. In this way the engine can use a lot of valve overlap, opening the intake early and closing the exhaust late, without losing too much fresh charge into the exhaust.

It takes time to scavenge the cylinder thoroughly and there is bound to be some residue of burnt gases which will dilute the fresh charge. But if the system is designed to make full use of the pressure pulse effects and of the inertia in the gases once they begin to move, then it is possible to increase the duration of the exhaust valve, leaving the closing point until after TDC.

As the intake valve opens before TDC, there is a lengthy period during which both valves are open. The undesirable effects of this are that burnt gas can mix with the intake gas, while fresh intake gas can be lost into the exhaust system. The desirable happening is for fresh intake gas to go into the cylinder and stay there while burnt exhaust gas is still leaving the cylinder and entering the exhaust system.

The probability of this happening by chance is very low but it can happen very effectively if the valve timing and the intake/exhaust dimensions are carefully matched to the engine speed. The effect is limited to a narrow speed range and is accompanied by bad effects at other speeds.

Other methods of increasing air flow and raising its pressure

There are other methods of increasing the air flow and raising its pressure, and some of the more popular variations are mentioned below.

(1) *Supercharger*: this is carried out by either a positive displacement or a centrifugal/axial blower, mechanically driven by the engine. The disadvantage of supercharging is that the types of superchargers available have been designed for larger-displacement engines, turning at lower speeds. they therefore tend to be bulky, difficult to install and drive, and may need a separate lubrication system. The engine may also need additional cooling, especially to the pistons and valves. Control of the fuelling is another difficulty, with the choice of blowing through a carburettor and pressurising the fuel system, or drawing through a carburettor which is then a long way from the engine and carries the risk of igniting fuel in the supercharger. Fuel injection controlled by micro-processor has eliminated most of the control difficulties. The advantage is a massive increase in torque which can be tailored to suit the application, giving boost in the low- and mid-ranges if necessary.

(2) *Turbocharger*: in this system a light, zero-displacement compressor is driven by an exhaust-powered turbine. Several versions small enough to match motorcycle engines are available but boost is only proportional to engine speed. They are compact and can be used with carburettors although fuel injection simplifies the fuelling, as with superchargers. The main disadvantage is the need to make the engine strong enough to withstand the extra thermal loads. Usually forged pistons are used, cooled by oil spray, with extra cooling for the cylinder head/valves along with special materials for the valves and springs. When the turbocharger is developed as original equipment the engines have proved reliable and further advances in materials, particularly in ceramics, will make smaller turbos practicable and will remove the problems caused by the increased heat flow. The high-speed turbines, turning at 200,000 rev/min need floating bearings plus a special oil supply and high specification lubricants.

Although they are not allowed in road racing, turbochargers are legal for the road and have to be the most effective way to produce large increases in long-term power. One problem is in matching the turbo to the engine, because exhaust turbines suitable for engines below 750 cc do not exist. Honda had to enlarge their CX500 V-twin to 650 cc when they turbocharged it and this was the absolute minimum, even with full factory development resources. The next problem is in fitting the turbine close to the exhaust and

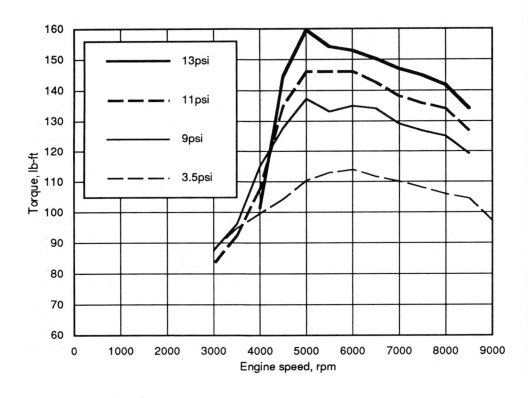
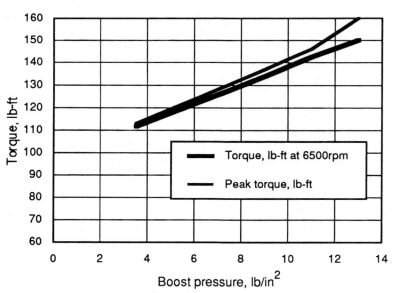

Fig. 1. The effect of turbocharger boost: a Rajay F40 fitted to a Suzuki GSX1100E with a 1498cc MTC big block. As long as the turbo is matched to the engine, torque is proportional to boost.

the compressor close to the intake so that losses are kept to a minimum – at least a V-twin lends itself to this, while most in-line fours create very awkward installation problems.

The steps for developing a typical application are:

Matching – choosing a compressor which will supply the necessary air flow and boost pressure over the required engine load/speed range. The turbine requires a certain minimum mass flow from the engine and the unit cannot begin to work until the engine has reached this region. If, as is usually the case, the turbocharger is meant for a bigger/more powerful engine then the turbine will not be driven up to working speed in the new engine's low- or mid-range. Practically no boost will be available until the engine gets close to peak torque; then, as boost becomes available, torque and gas flow through the engine increase. More mass flow drives the turbine to a more efficient speed region and the compressor can now deliver more air flow at higher boost pressure; engine torque and speed go up further. At best this kind of mismatch will make the engine peaky. The midrange will be worse than the standard engine because of the restriction of the turbine and the lengthy intake; at higher revs the power increase will accelerate upwards. At worst, this kind of system will only make a useful increase in power close to the engine's maximum safe speed. Therefore the turbine must be able to respond to the engine's low- to mid-range exhaust flow.

A/R ratio – this is the geometry of the turbine housing, where A is the area of the turbine inlet at its narrowest point and R is the radius of this point, that is the distance of the centre of the area A to the centre of the turbine's shaft. The A/R ratio dictates the turbine speed for a given amount of exhaust gas energy. If the A/R value is larger, the turbine will turn slower, if it is smaller, it will turn faster, for the same amount of exhaust gas flow.

Surge – the compressor has to be capable of supplying more air than the engine needs, otherwise it would not be able to pressurize the air to provide 'boost'. If the engine takes away air as fast as the compressor can deliver it, this is just inefficient and there are no significant power gains. If the compressor is allowed to build up pressure, then its blades make increasing contact with the air flow and it enters a region of peak efficiency, depending on the speed of the air relative to the speed of the compressor blades. This region of air flow versus pressure varies with the design of the compressor. As the pressure increases further, the air flow reaches a condition known as 'surge', in which it stalls, that is the compressor blades no longer control it and the flow can reverse direction. This leads to violent changes of gas flow and can do a lot of damage. As surge is associated with excess pressure for the rate of air flow, it usually occurs when the compressor is too big for the engine.

Fuelling – the first choice is fuel injection, with the injectors close to the intake valves so that only air is drawn through the compressor. The second choice is to have a single carburettor upstream of the compressor, usually with a large chamber between the compressor and the intakes to the individual cylinders.

Engine modification – the engine will not be able to make use of exhaust or intake effects, so late valve timing and a lot of intake/exhaust overlap are not necessary. If the compressor is well-matched to the engine characteristics then it will provide boost at low engine speeds and the trapping efficiency will be improved by having short duration cams and by maximizing valve area. Forged pistons are essential to cope with the increased heat flow and as they have to be changed it would be as well to consider using bigger bores. Making the engine bigger would make it easier to match the turbine and it would leave room for larger valves as well. As high compression pressures will be available from the boost control, it is often necessary to lower the engine's compression ratio to reduce the risk of knock. Extra piston cooling is available either by spraying oil through jets on to the undersides of the pistons, or if there isn't a convenient pressure source for this, by increasing the side clearance of the big end bearings in order to increase the oil splash feed to the cylinder walls and pistons.

The bottom end of supercharged engines usually has an easier time than the rods and crankshafts in naturally-aspirated machines. This is because the extra charge density tends to cushion the piston at the top of the stroke and although the engine makes more torque, the build-up to peak pressure is not so violent. Turbocharged engines tend to run lower peak speeds, so inertia forces are less. As long as the motor can stand the new torque levels then rods, main bearings and big end bearings should not be a problem. Motors with pressed-up, roller bearing cranks (e.g. Suzuki GSX models) usually have the crank pins welded to the webs and high tensile studs are usually used to hold the cylinder and head to the crankcases.

With the increase in charge density and thermal loading, the ignition requirements will be greater and a more powerful system will be needed, typically something like the American Dyna S, with an MSD multiple spark amplifier. The ignition timing has to be optimized (see Chapters 8 and 11) but the biggest danger is from detonation which, when engine output is raised by 50% to 100%, can destroy pistons in a matter of seconds. Therefore a fairly conservative – retarded – ignition curve is used, sometimes in conjunction with water or water-alcohol injection. A knock sensor can also be used with electronic ignition or fuel injection control, to retard the timing and/or richen the mixture whenever detonation is sensed. (Detonation causes quite powerful shock waves to be transmitted through the engine. Knock sensors are like set screws with piezo-electric crystals embedded in them and are screwed into the head or cylinder block. Shock waves caused by detonation compress the crystals, generating an electric signal which is detected by the engine management system.)

One other safeguard is needed, especially if fuel is mixed with the air passing through the turbine, and this is a stay to prevent a backfire blowing the intake manifold away from the cylinder head.

Turbochargers run at very high speed, 150,000 to 200,000 rpm is usual for those sized to work with motorcycle engines. They usually run plain, 'floating' bearings – that is, one plain bush on top of another, with fairly generous radial clearances, thrust washers at either end and labyrinth or piston ring type seals. High grade oil is necessary (usually a multigrade 5W-40 or 10W-40, to API classification SF or SG) to provide suitable flow in all conditions and to cope with the high temperature in the exhaust side of the turbine. As the turbine will continue to spin for some time after the engine has been shut down, a pressure feed to a small reservoir above the supply to the bearings is necessary.

Wastegate – this is a valve which allows exhaust gas to by-pass the turbine and is operated by a pressure-sensitive mechanism in the intake. When the boost pressure reaches a pre-determined level, the valve opens and exhaust gas is diverted from the turbine until the boost pressure falls to a safe level. Usually the mechanism is adjustable.

(3) *Exhaust energy converter*: in theory, this is any device which can take energy from the exhaust and re-cycle it to give useful power. In practice the only device to appear so far is Comprex developed by Brown Boveri. This consists of a rotating drum, full of axial cells which are open at one end to the exhaust and at the other to the intake. The brief inter-mixing which takes place is able to transfer energy from the exhaust to the intake without any physical dilution problems. The disadvantages for motorcycles are those of bulk and the need for an extra drive, while materials development has not yet made it a practical prospect for SI engines.

(4) *Different fuels*: (a) fuels like methanol which have less heat content than petrol but have much greater knock resistance. This allows very high compression ratios to be used, with a predictable increase in thermal efficiency. Because the heat content is low, more fuel is needed and the greater latent heat of evaporation allows the engine to run very cool. The machine needs special materials in the fuel lines, tank and carburettor and the fuels are not legal for street use in the UK, although some countries are experimenting with alcohol-based fuels and it is likely that additives derived from a similar source will be used in pump fuels.

(b) Fuels which liberate oxygen, such as nitro-methane. The standard description of nitro-methane is that it does not add power, it multiplies it. The fuel actually gives off oxygen which can be used to burn more fuel,

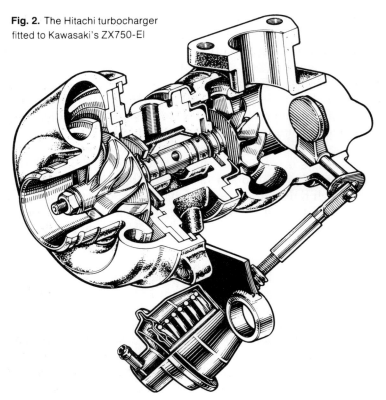

Fig. 2. The Hitachi turbocharger fitted to Kawasaki's ZX750-E1

consequently an immense amount of fuel can be burnt per cycle. The disadvantages are the cost, the difficulty of handling and the development costs. The rewards are spectacular. Illegal for practically everything except drag racing.

(5) *Separate oxydizing agent*: a typical installation is a bottle of nitrous oxide, stored in liquid form under pressure of about 800 lbf/in^2. A valve, usually operated by a lever or a solenoid connected to a control switch, introduces the nitrous oxide through a nozzle downstream of the carburettor. Additional fuel is also supplied via the same or a separate nozzle, hopefully with a failsafe valve which shuts down the N_2O if the fuel supply fails. The liquid vaporizes instantly, substantially cooling the intake stream and the engine components. It gives off oxygen which is then available to burn the extra fuel, with power limits only dictated by the melting point of the engine. The disadvantage is the limited supply (a reasonably-sized bottle lasts less than 60 seconds on a 100-bhp engine) and the control, which is either on or off, although it can be staged. It is legal for road use and drag racing but not for road racing.

(6) *Octane boosters*: high octane fuel permits the use of higher compression ratios which give more power. While pump petrol is restricted to 86–87

octane (MON), Avgas is available at 100–115 MON octane, although it cannot be legally used in road vehicles. Octane boosters can be added to petrol to raise its octane rating; most of these are toxic and are prohibited in road race engines.

Special high octane fuels (for example Elf Bluegas) are available and some of the oil companies will make small quantities of fuel to a given specification. However the trend among race organisers is to insist that unleaded fuel be used and that it should be generally available as 'pump' petrol.

Table 1.1 Fuel octane ratings

Fuel	MON	RON
4-star (BS4040)	86min	97min
Unleaded premium (BS7070)	85min	95min
Super unleaded	87	98
Shell racing unleaded	88.8	99.9
4-star (measured)	86.8	-
with various octane boosters	86.6-89.7	
Avgas 100LL	99.5min	
Elf Moto 119 (Bluegas)	>110	>119

Note: MON, RON = motor octane number, research octane number.

Table 1.1 gives octane ratings of various types of fuel. The fuel is compared, in a single-cylinder variable-compression CFR test engine, to a blend of iso-octane (which is very knock resistant) and n-heptane (which has little knock resistance). A figure of 87 means it has the same resistance as an 87% octane blend. The MON test is run at higher speed and is considered more realistic. Figures over 100 are achieved by adding tetra-ethyl lead (TEL) to a 100-octane reference fuel. This is the 'lead' of leaded and unleaded gasoline, not the dark grey stuff found on roofs and windows. When it passes through engines lead bromides appear, which form the once-familiar light grey deposits inside exhaust pipes. Unleaded fuel usually has oxygenates – alcohols, MTBE etc. – instead of lead. These leave black, sooty deposits in the exhaust. It doesn't signify a rich mixture. Some octane boosters work – these tend to contain aniline which is carcinogenic and as nasty as the TEL which it replaces. Oxygenates work less well and some boosters actually do nothing or reduce octane ratings. If the fuel already contains the same booster, it won't make much difference; mixing different types of booster makes the biggest difference. Although the LL in Avgas 100LL stands for Low Lead, the Ministry of Defence only mean this in relation to other Avgas grades – 100LL is recognizable by its blue dye (not to be confused with blue gas which is actually yellow!) and contains 0.56 gPb/L compared to leaded four-star which contains a maximum of 0.15 and unleaded premium which contains less than 0.013. Elf's Moto 119 was actually designed for very high revving two-strokes but it gives an idea of what is available. Its lead content is twice as high as Avgas LL.

13

Losses

Reducing losses has the same effect as increasing power; in fact it is a wasted effort to raise the power output if the engine's losses have not been cut to an absolute minimum.

There are two basic types, friction (or mechanical) losses and pumping losses which are those concerned with moving the gas through the engine, displacing gas inside the crankcase and in driving ancillary equipment such as electrical generators, oil pumps, cooling system pumps and fans, etc.

(1) *Friction:* the engine must be set up as effectively as possible by blueprinting and modifying bearings/clearances to suit its new specification (see Chapter 10). Improved lubrication may help and it may be possible to modify parts to reduce side-thrust, e.g., using a longer connecting rod, using straight-cut gears instead of helical gears. Oil drag can be reduced by fitting baffles in the sump, or using a dry-sump kit. Better lubricants and more efficient oil control may also reduce frictional losses.

(2) *Pumping losses:* an increase in air flow will increase pumping losses and, apart from making the ports, etc, flow as easily as possible, there is not much that can be done about this. However, crankcase pumping can be alleviated by increasing the breather capacity and ancillary equipment can either be removed or it can be replaced by specialized parts.

(3) *Reciprocating parts:* the force necessary to accelerate each moving part, and then to slow it down again, can only come the engine's power stroke. As the force is proportional to the mass of the component, the losses will be reduced if any of these parts can be lightened.

Performance Bikes ran a series of motoring tests on a Kawasaki GPX750, using a Laurence-Scott DC dynamometer at MIRA. In this the dyno was used like an electric motor, to turn the engine at various speeds, while the torque required was measured. The engine oil was kept at a constant temperature and the motor was run complete (apart from carburettors and exhaust system); then the head was removed (eliminating the main pumping losses and the drive to the valve train); next, the pistons and connecting rods were removed (the major source of friction in the engine); finally the alternator was removed (leaving the main bearings, oil pump and the transmission). The power losses caused by these various assemblies are shown in Fig. 3.

Honda's R & D company made an analysis of 145 production four-strokes (single, twin and four cylinder) in which they made theoretical calculations of the frictional losses and then compared this with tests made on actual machines.

They came to the conclusion, backed up by dyno measurements, that most engine friction came from the pistons and crank journals and that losses in the valve train and transmission were small in comparison. In particular, they noted that:

all friction	∝	$\sqrt{\mu}$
	∝	$1/b$
	∝	$\sqrt{s}$
	∝	$\sqrt{rpm}$
except:		
crankpin	∝	$(pin\ \varnothing)^{3/2}$
	∝	$\sqrt{(pin\ width)}$
	∝	$1/s$
	∝	$(rpm)^{3/2}$
main bearings	∝	$(journal\ \varnothing)^{3/2}$

where:
b = bore
s = stroke
rpm = engine speed
⌀ = diameter
μ = oil viscosity

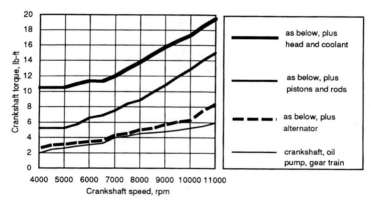

Fig. 3. Pumping and friction losses measured on a Kawasaki GPX750, expressed as torque at the crankshaft, and as a percentage of the total power.

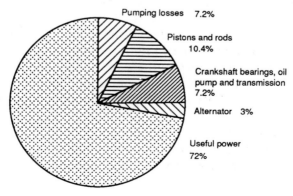

Chapter 2
Tools and equipment

The type of modifications made to four-stroke engines demand some very specialized tools and equipment. The ability to measure components accurately is essential and there will be occasions when it is necessary to machine them with a high degree of precision.

Some tolerances are critical (see Chapter 10), others are just difficult to measure; in these cases the manufacturers often use selective assembly, in which new components are measured accurately at the factory and are then stamped with a numeric or colour code. Tables supplied in the workshop manual will then give the code needed for the mating part. So the first item of equipment is the manufacturer's shop manual, which will give all the baseline data for the machine and provide information on selective assemblies.

The problems arise when these assemblies are modified, or when the parts are worn (such as the shims used to adjust valve clearances). In this case it will be necessary to measure the part accurately; it may also be necessary to measure the stock parts in order to find the actual dimension instead of the coded version.

Consequently the second piece of equipment will be your own manual – a file built up containing your own measurements (and methods of measurement where these are different from those used by the factory). This can be a simple notepad, although a loose-leaf ring-binder is more adaptable.

As work progresses on the engine, the file can be extended to include details of tests and their results and will build up into a full specification for the engine. This makes it simple to repeat the same work on another engine, to go back to an earlier specification, or to compare the results of one test with another. To make it easier to use the notes, a means of cross-referencing items should be devised and as many diagrams as possible be used. It is amazing how a page of numbers which was simplicity itself when first written, becomes incomprehensible a few weeks later.

Further requirements for carrying out an engine overhaul are to have the right tools including any special items of equipment, and to have the ability to do the work. Mechanical skills and aptitudes vary but as a large part of the work hinges on competent engine building, the success and reliability of any tuning work cannot be taken for granted. The measured change in power output between a new engine and the same motor after running-in can be in the region of 5 to 10 per cent. The same applies to a blueprinted engine – depending on how close to tolerance it was in the first place. It follows that inaccurate assembly work can also cost the same amount of power.

This particularly applies to measurements made with precision tools like micrometers. Given a simple metal bar to measure, six people who had never used a micrometer before would probably come up with six different answers. Given something like a piston, which looks round and parallel but is actually oval and tapered, and the chances of getting the same answer twice are very remote.

With any unfamiliar equipment, make sure you understand how to use it and then practice until you get consistent results. It is also necessary to keep an overall understanding of what you are doing and why you are doing it – although it sounds obvious, it is very easy to get totally absorbed in minute details and lose track of the real object. For example, a piston measured beside a warm kitchen stove will have a significantly different size to when measured in a freezing workshop. And a cylinder measured immediately after boring will have a different diameter after it has cooled down. Any mating parts must always be measured together, at the same time and place.

Measuring equipment

1. Vernier caliper, with depth gauge. The most versatile, general-purpose piece. A good one will measure to 0.02 mm.
2. Micrometer – external. Accurate to 0.01 mm, this is essential for measuring bearings, pistons, etc. They are usually sized 0–25 mm, 25–50 mm, 50–75 mm and so on, which means that at least two and possibly three will be needed. A standard test piece is included so that the accuracy can be checked and adjusted if necessary.
3. Micrometer – internal, depth type, etc. Several variations are available for different applications but unless one type of measurement is made frequently enough to justify the cost of the equipment, it is usually enough to use a suitable bore gauge for inside measurements, used in conjunction with an external micrometer.
4. Dial gauge plus stand and magnetic clamp. Accurate to 0.01 mm and available with various full deflection lengths – make sure there is enough travel for your purpose. Stems with differently shaped followers are available which simply screw on to the base stem. This is an essential instrument for checking valve lift, measuring shaft runout/truth, endfloat and backlash in gears. It must have a totally rigid stand and clamp, and must be positioned so that the stem is either parallel with or at right-angles to the part being measured, otherwise there will be a vector error. (Also called a dial indicator or comparator.)
5. Degree disc. Useful for setting cam positions, measuring valve lift, etc. It must be rigid and mounted concentrically on the shaft – which will

probably need a special fitting for each type of engine; a 'universal' type usually wastes time and is less accurate because of the inconsistency of the mounting. The pointer – usually a piece of welding rod – should be securely mounted and bent so that it is rigid and close to the face of the disc to avoid any parallax errors.

6. Feeler gauges (thickness or wire gauges). If a large amount of work is to be carried out on a particular engine it is worth adapting a set of feelers to suit it; for instance bending the appropriate gauge into an L or a Z shape so that it can be slid easily into position, e.g. between cam and follower. Similarly, two gauges can be made into a go/no-go tool to facilitate service checks. If the tolerances for a clearance are 0.09 and 0.11 mm then two feelers of these thicknesses can be incorporated so that (if the clearance is within the tolerance) one end of the tool will go into the gap while the other will not.

7. Plastigage. For measuring bearing clearances. This is a material not unlike plasticene which is formed in thin, accurately made strips. A piece is placed between bearing surfaces and the bearing assembled to its usual torque or tension, squashing the Plastigage. The width of the squashed strip, used in conjunction with a scale provided in the kit, gives the bearing clearance. The bearing surfaces must be dry, the strip must be placed parallel to or at right-angles to the axis of the journal, it must be kept clear of any oil way and the assembled bearing must not be moved.

8. Electrical tester. This is becoming more important in evaluating ignition circuits. It should have an ohms scale (not kilo-ohms) so that pulser units and coils can be tested.

9. Burette. Essential for measuring volumes of combustion chambers, piston crowns, ports (to compare one with another), etc.

10. Fuel level gauge. Two types are available, although both can be made up quite easily. One is simply a depth gauge made to fit the carburettor float-bowl face and shaped to clear the floats. The other is a tube which, when connected to the float-bowl drain plug, shows the actual height of fuel in the float bowl. Transparent float bowls are available for some carburettors.

11. Vacuum gauge. Useful for synchronizing multi-carb engines; easier to use if there is one gauge per cylinder. Mercury columns are probably the most reliable. A vacuum gauge can also be used to set part-throttle positions either to run mixture loops on a dynamometer or to compare road loads with dyno loads.

12. Compression tester. Useful for comparing individual cylinders as well as detecting faults.

13. Oil pressure tester. Will need adaptor to suit particular engines. May be useful if the lubricating system has been modified in any way.

14. Stroboscope. Possibly the best way, sometimes the only way to check

ignition timing and advance curve characteristics. Note that the marks should be checked by using a dial gauge on top of the piston.
15. Surface plate, V-blocks, mandrels. Essential for checking dimensions and truth of components such as crank assemblies, connecting rods and so on. Specialist machine shops should have this facility.

Modification

The equipment falls into two categories: hand tools and machine tools. Unfortunately it is the latter which come in for most use when four-strokes are extensively modified. This means that much of the work is expensive, either because of the cost of the tools or because the work has to be sent out to a machine shop.

Hand tools: a selection of small files and rotary files with a flexible drive will be enough for work on cylinder heads and ports. By far the best type of drive is the high-speed motor (turning to something like 20,000 rev/min) used by die-makers, etc. The alternative, a motor turning at 3,000 rev/min, even with a 2:1 step-up gear, will make the tools hard to control and will give flexible drives a short life.

Other tools likely to be used are valve seat cutters, able to cut at 45 degrees (or whatever the valve seat angle is) plus wider and narrower angle cutters to trim the seat width. A single-point cutter is the easiest to use if the valve seats have to be cut back a long way.

Finally, an adjustable hone to set the cylinder to piston clearance and a set of reamers for parts like small-end bushes and valve guides may also be needed.

Machine tools: at fairly regular intervals a wide variety of machine tools will be needed and because of the precise nature of the work there is no substitute, apart from buying ready-made components if such items are available.

A lathe is used for a whole variety of nondescript turning jobs and a mill is needed for basic modifications such as machining cylinder heads or cylinder blocks, increasing the valve cut-aways in piston crowns and so on. A boring bar is also quite useful although there is a lot to be said against using the small types which simply centre on the existing bore. It is better to go to a specialist who has the equipment to bore all the cylinders parallel and square to the axis of the crankshaft.

Polishing and shot-peening processes are useful where parts are to be stress-relieved or where forged parts have been modified and need to be re-hardened. Again, a specialist firm should be used and it is essential that they use a process designed to shot-peen the surface to toughen it and not simply a shot-blast cleaning process (see Appendix).

Finally a cam grinder, plus the ability to build parts up by welding or metal spraying is often a preferable activity to simply buying someone else's cams.

In all of these operations, if you do not have the skills or facilities to do this work yourself, it is important to talk to the people who will be doing the work, partly to evaluate their skills and partly to make sure that they know exactly what is wanted and you know exactly what they require. You could not expect a machine shop to dismantle parts and, where bearing surfaces or oilways need to be masked off, you can be sure of the job if you either do it yourself or at least find out how they intend to do it.

Testing

Once again, the equipment needed to test engines is both expensive and bulky. The main items are an air flow rig, a dynamometer and fuel flow meters.

An air flow meter, as shown in Fig. 4, uses a pump to draw air through the item under test and through a chamber with a standard orifice. The measurements compare the pressure drop across the standard orifice against the pressure drop across the component. It is useful for comparing similar items, to see which will flow more under the same conditions, or to make sure that matching parts have equal flow characteristics. It is worth bearing in mind that increases in flow as measured by the flow rig rarely show up as proportionate increases in power when the parts are fitted to an engine. This is partly because the flow rig can only evaluate steady-state conditions and cannot simulate the sometimes violent pulsations in the real flow of air through an engine. It is also partly due to misinterpretation of the results; some assumptions have to be made (such as the pressure drop caused by the engine) and it only evaluates the component in isolation. To take an extreme case, it will not matter how much work is done to the valves and the ports if the throttle valve is never opened.

Any modified parts have to be considered in conjunction with the rest of the engine and in their actual operating conditions. Where the air flow rig is of help is in providing data for use with other components. It will show how the air flow increases as a valve is progressively opened, and from this the maximum lift of the valve can be determined. This information can then be used to design the cam profile and to calculate the effective time-area of the valve (see Appendix).

A dynamometer is the most useful tool of all, as long as its limitations are recognized. In essence it contains a rotor which is driven by the engine and to which a variable load can be applied. This may be a load transmitted by an impeller immersed in water (as in the Heenan & Froude water brake shown in Fig. 6), or it could be loaded by increasing the current in an electrical generator, or by some other means of transmitting an indirect drive.

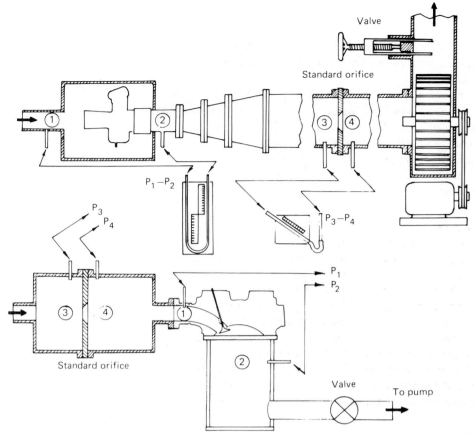

Fig. 4. Typical layouts used in air flow rigs. The valve is used to adjust the flow until a certain pressure drop appears across the component, e.g. 10 inches of water and then the volume flow rate through the orifice is taken from a formula such as:

$$Q = C_d C_a a_0 [1 - (a_0/a_3)^2]^{-1/2} [2(p_3 - p_4)/d]^{1/2}$$

where Q is the volume flow
 C_d, C_a are the discharge and area coefficients of the orifice
 a_0 is the diameter of the orifice
 a_3 is the diameter where p_3 is measured
 p_3, p_4 are the upstream and downstream pressure measurements
 d is the density of the air (which may be subject to various correction factors for temperature, pressure and compressibility at high speeds).

The formula is usually simplified to something like

$$Q = K[(p_3 - p_4)P/T]^{1/2}$$

where K is a constant for the rig, and
 P, T are the ambient pressure and absolute temperature.
 The air flow coefficient for the object under test is

$$C = \frac{a_0 C_0}{a} [(p_3 - p_4)/(p_1 - p_2)]^{1/2}$$

where C is the air flow coefficient of the component
 a is the diameter of the component
 p_1, p_2 are the upstream and downstream pressure measurements across the component.
 Dimensions of the standard orifices are contained in a British Standard

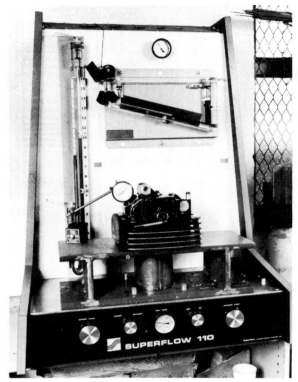

Fig. 5. A four-valve Honda cylinder head set up on a Superflow air flow rig. The manometer on the left is recording the pressure drop across the valve, while the inclined manometer records the pressure drop across the standard orifice, from which the air flow is calculated.

Against whatever type of loading is used, the rotor attempts to drive the stator, which is free to pivot but is restrained from moving by either a load cell or an arm attached to a weight and a spring balance.

The force generated is recorded by the load cell or spring balance, and this, multiplied by the length of the connecting arm represents the torque on the stator. A speed sensor also measures the speed of the dynamometer shaft and this speed, multiplied by the torque, multiplied by a suitable constant (called the brake constant) gives the horsepower being absorbed by the dynamometer.

The load can be varied (by sliding sluice gates between the impeller and stator, or by varying the field coil current) and the engine, with its throttle wide open (WOT) will accelerate to a speed where it is balanced by the load, run until it reaches a steady state and then the speed and load readings can be taken.

Because horsepower is a function of torque and speed, it is not altered by gear ratios (ignoring the frictional losses caused by the gears). However, torque is amplified by reduction gears in the same way that speed is reduced.

If the power produced at the crankshaft is P and the transmission losses amount to L, then the power received and measured at the dynamometer will be:

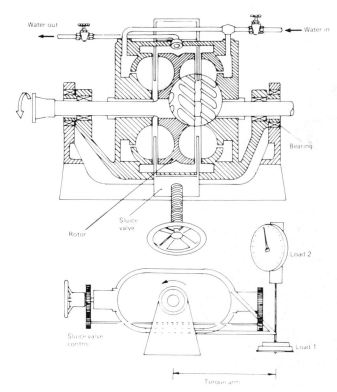

Fig. 6. Operating principles of the Heenan & Froude dynamometer. The drive shaft runs in bearings and glands in the dyno housing, while the whole structure pivots in outer bearings held by the fixed bed. Therefore frictional 'losses' in the main bearings and glands are included in the total, measured torque

Fig. 7. Kawasaki engine installed on a Schenck U1–16H dynamometer

$$P - L = T_d \cdot NK/G$$

where T_d = dyno torque
 N = crank speed
 G = reduction gear
 K = brake constant
 (N/G = dyno speed)

crankshaft torque T_e will be given by:

$$T_e = T_d/G$$

so

$$T_e = (P - L)/NK$$

This is the effective torque at the crank as it incorporates transmission losses (L). The actual crank torque will be P/NK.

At each position the fuel flow to the engine can be measured using a flow meter and this can be calibrated in volume flow (pt/h) or in mass flow (lb/h). Dividing the fuel flow by the power obtained, gives the specific fuel consumption (SFC), measured in pt/hp-h or lb/hp-h (or gm/kW-h). These three values; power, torque and SFC are plotted against crank speed to give the characteristic WOT curves for the engine.

There are several dyno configurations with various types of control:

1. Water brake, as described above (Heenan & Froude DPX models, Schenck U1-16H, Superflow 800 and 901). The load–speed parameters must match the engine characteristics but within this region this type of dyno gives very accurate, repeatable results. They are good for holding steady-state conditions, for example to set carburation. The manual control is slow which makes it difficult to 'catch' unstable engines or those with very peaky power characteristics. Some have motorized controls for this reason. The Superflow has automatic computer control which allows it to take readings at selected speed intervals, from a pre-set start speed to a pre-set maximum speed, holding the engine at each speed for a pre-set period. The results are logged by sensors and stored in a computer file. This is very useful for testing engines in transient conditions and for running engines which are unstable or extremely highly tuned.

The drive is usually taken from the gearbox output shaft, using either a chain or a female spline (replacing the gearbox sprocket) with a shaft and universal joint coupling to the dyno. Some installations will accommodate the whole machine, others need the engine to be removed.

2. Eddy current dyno (Froude G4, Bosch FLA203). Here the load is supplied by driving an electric generator whose output is fed back into the field coils to resist the motion. As the energy is frittered away in the windings, they get

very hot and need water cooling. Manual control is by altering the current which creates the load but there are also two semi-automatic modes. The dyno can be set to maintain a specified load (while the dyno or engine speed is monitored) or it can be set to hold a steady speed (while the load is monitored). These can be very useful in tests where a third variable (such as turbo boost or engine temperature) is introduced, for example, see Fig. 57. A fully automatic mode can vary load and speed in a predetermined way, usually to simulate road loads.

The installation is similar to water brakes and the engine must match the load–speed characteristics.

The Bosch type is driven by a road wheel roller – see (4) below.

3. Direct current or swinging field dyno (Laurence-Scott). Similar to (2) but uses a DC machine which can be driven by the engine against a load created by feeding power to the field coil, or as a motor which can drive the engine. This can be used for motoring tests, to measure engine friction and pumping losses.

4. Inertia dyno (Bosch LPS002 and FLA203, Dynojet model 100). Both of these types have a roller driven by the motorcycle's rear wheel and the roller is connected to a very heavy flywheel (one tonne in the Bosch, 1000 lb in the Dynojet).

While the engine is accelerated on WOT, sensors record the speed of the flywheel in real time and the data is stored in a computer. Software then calculates the flywheel's acceleration and, as its moment of inertia is known, can calculate the torque required and hence the horsepower. Road wheel speed is known and another sensor can pick up signals from the ignition to monitor engine speed. All the data is stored in a computer file. This type cannot hold the engine at steady speed but it can monitor deceleration when the throttle is closed and so produce reverse-thrust data when the engine is on the overrun and, if the drive to the engine is disconnected, the power absorbed in the transmission. There are a number of disadvantages but despite this, inertia dynos can give repeatable results and are good as comparators, with the big advantage of taking minimal time to set up each test.

The Bosch FLA combines the inertia of the heavy flywheel with the load control of an eddy current dyno.

The disadvantages are:

(a) if the rear of the bike is strapped down too tightly there is a risk of exceeding the tyre's load/speed index and damaging its carcass. This risk is increased considerably by the type of dyno that uses twin rollers and for this reason a separate tyre should be kept exclusively for dyno testing.

(b) There will be some slip between tyre and roller, increasing when the bike is not strapped down tightly and when the power and wheel speed increase. The twin roller Bosch dynos can monitor slip because there are speed sensors on both rollers but only the front one is loaded.

(c) Unlike other types (where the load is held in a swinging frame so that friction in the dyno bearings is added to the measured load) friction in the roller bearings and windage losses in the roller/flywheel are not accounted for unless the software makes an allowance for them. This could be considerable at high speed (roller speeds approaching 200 mph are not uncommon) and it could change with different ambient conditions, wear, etc. This type of dyno does show slightly differing torque curves when the engine is run in different gears. The readings would also vary slightly if the rotating inertia of the wheel/driveline were altered, but this would presumably be small compared to the size of the flywheel.

(d) The engine cannot be held at a steady speed under load if the dyno is a pure inertia type.

The reason that steady speeds are desirable is that the engine takes a few seconds to stabilize (although the argument against this is that it wouldn't be allowed to stabilize on the road or track, so transient conditions are more realistic) and fuel flow meters take several seconds to reach a steady reading. Flow meters can take the form of small turbines inserted into the fuel line, with Hall effect or magnetic sensors sending signals to a meter or to be stored in memory. These follow fluctuations in fuel flow fairly quickly, but they are inevitably some distance upstream of the engine. Other types (for example, Rotameter) have metal spinners which are lifted in a sight glass by the fuel flow and can take many seconds to stabilize. CO meters fitted in the exhaust stream also take five or six seconds to respond fully.

(e) If the engine fails mechanically then the inertia of a flywheel (or of the generator rotor in an eddy current dyno) will continue to drive the engine, possibly doing more damage, until the operator disengages the clutch. Water brakes simply stop when engine power is lost.

Chapter 3
Development

The first stage in engine development is to evaluate the engine against other available machines and to see if there are any compatible parts used on larger versions or competition variants. Race shops or accessory firms may also offer high performance parts. For competition use much can be said in favour of choosing the engine which already gives the most power but we also have to accept reality and acknowledge that the reason for tuning is often to uprate an uncompetitive model.

There will inevitably be restrictions in the form of competition formulae or the legal requirements governing road machines, and these must be taken into account at an early stage. One of the first requirements is the rule book.

Four-stroke motors are available in many different configurations, each with its own advantages and problems.

1. Single cylinder
In its favour, the single is simple, narrow, there are special race formulae just for single-cylinder machines, and it is easy to develop. Factories often use a single cylinder from a multi to do basic development and it is not unusual to find, say, a small trail bike engine which is, in essence, one-quarter of an in-line four. Sometimes parts and development work can be swapped between models like this, so it is always worth examining other models in the factory line-up which have the same bore and stroke or other similarities.

The disadvantages are a lack of piston and valve area for the engine size, a heavy piston which in turn leads to vibration, high stress/low rev limit, and relatively heavy engine castings and mountings. Many singles now have balance shafts which effectively deal with vibration and mounting problems.

2. Parallel twin
The 360-degree twin is just a more convenient format for a single; it is simply a very over-square single in most respects. A 500 cc twin compared to a 500 cc single of conventional proportions would have more valve and piston area, would not be so tall, but would still have the same disadvantage of vibration.

The 180-degree twin eliminates some vibration at the expense of uneven firing intervals. Otherwise its proportions are the same as the 360-degree twin.

Fig. 8. Norton's P86 parallel twin prototype, built by Cosworth. The engine gave 90 to 100 bhp, making it one of the most powerful parallel twins ever

Parallel twins can run very smoothly when balancer shafts are used, but if a big bore kit is fitted, it will be necessary to modify the balancer to allow for the extra piston weight.

3. V-twin

This has most of the advantages of a single, coupled with less height, better piston and valve area and potentially perfect primary balance. The 90-degree V has complete primary balance, running both big-ends on one crankpin but the 90-degree angle between the cylinders can make the engine rather unwieldy and does not lend itself to a tidy carburettor and exhaust layout.

More compact, narrow angle V engines can give total primary balance but they need two crankpins and the pins must be offset by an angle X where

$$X = 180 - 2V$$

where V is the angle between the cylinders.

Fig. 9. The narrow, compact layout of Ducati's 600 and 750 cc V-twins made them successful in F1 and F2 racing

4. Flat-twin

A horizontally-opposed twin with the crankpins set 180-degrees apart has perfect primary and secondary balance; the only vibration will be caused by a small rocking couple which is set up because one cylinder is slightly in front of the other. While lacking in height its width/length causes installation problems.

5. In-line 3

There is a lot to be said for these engines in theory but in practice they have always been disappointing. With 120-degree cranks they have perfect primary and secondary balance, but the rocking couple acting over the full length of the crankshaft always causes vibration. In other respects – size, piston area, valve area – they split the difference between twins and in-line fours. When allowance has been made for the difficulty of manufacturing a 120-degree crank, the extra complication of the four-cylinder engine is not a great problem.

The 180-degree crank makes the triple cheaper to build than a four, but it then suffers similar vibration problems to a 180-degree twin.

Fig. 10. One of the most compact versions of an in-line four, based on the Yamaha FZ750. On this machine, the layout of the engine and liquid cooling system has enabled them to be squeezed into the most suitable shape

6. In-line 4

Practically every engine manufacturer in the world makes an in-line four, which tends to suggest that the layout is the best compromise between maximum piston/valve area, minimum piston mass, manufacturing cost, overall dimensions and balance. In most ways, it is the least that has to be done to get acceptable power. Balance is as important to manufacturers as to race engine builders because it means that lighter castings and lighter engine mounts can be used, while cycle parts and controls will not have to be isolated. An in-line four runs like two 180-degree twins, it has inherent primary balance and the rocking couples cancel one another out, leaving the secondary out-of-balance force, whose frequency is twice engine speed, to be absorbed through the frame. A few models, for example the Honda CBR1000, have balance shafts turning at twice engine speed in order to take out primary and secondary vibration.

The crankshaft is fairly long, especially if ancillaries such as the generator are hung on the end of it, and this makes the engine fairly wide. It can also induce torsional vibration problems, in which the load on the crank makes it twist and then spring back. The degree of twist depends upon the load and the strength of the shaft (or its length to thickness ratio). The rate at which it springs back depends on its natural frequency, which will be in proportion to the thickness of material. Problems occur when the engine frequency coincides with the natural frequency. If the shaft is operated at resonance it will break very quickly. The manufacturer will counter this by making the shaft thicker, raising its natural frequency well out of the engine's operating

Fig. 11. This V4 Honda engine is used in a variety of models. The power unit is narrow, short, has has good balance and plenty of piston/valve area. The race (and later road) versions have gear driven camshafts

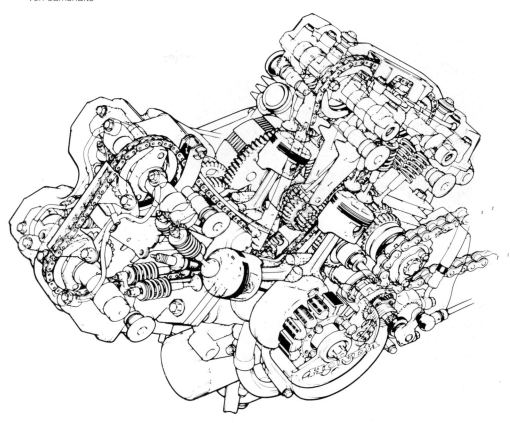

zone. He will also reduce the torsional stress by minimizing the weight carried on the ends of the shaft and by taking power off one of the centre webs, not from the end. However, a tuned engine may well be brought back into the danger zone, because the engine speed will be raised and the crank may be lightened, lowering its natural frequency. Also heavier loading will increase the stress on the shaft and may cause fatigue failure.

With layouts more complicated than in-line fours, the routing and performance of exhaust and intake systems becomes a problem.

7. V4

Although more expensive to make than an in-line engine, the V4 is more compact (some 20 per cent shorter in the crank, not as tall but trickier to install) and has complete primary balance. When in-line engines reached the levels of power that demanded liquid-cooling, the V4 then became a practical proposition. In theory, the V4 (or a square or flat four, or a V6) should be the ideal motorcycle format.

Fig. 12(a). Honda's prototype V4 NR500 had oval bores and pistons, giving it many of the advantages of a V8 in a class which only permitted four cylinders

Fig. 12(b). Honda continued oval-piston development and produced the NR750 prototype which used these components

8. Square 4
This has roughly the same advantages and disadvantages as the V4 except that it needs two separate cranks.

9. Flat 4
The flat 4 has perfect balance, but as with the horizontally-opposed twin, its shape does not lend itself to motorcycle installation.

10. In-line 5
With 72-degree cranks, the in-line 5 has complete primary and secondary balance, but, as with the in-line three, there is a large rocking couple built up over the length of the (long) crankshaft. Although it has more piston/valve area than a four-cylinder engine, the overall width is likely to be a problem, coupled with a greater susceptibility to torsional vibration in the crankshaft.

11. In-line 6
With 120-degree cranks, the in-line 6 gives the same balance as two triples; complete primary and secondary balance and the rocking couples cancel one another out, so the engine is completely smooth and has the added advantage of evenly-spread firing intervals. With more piston/valve area than a four-cylinder engine, the six has a lot going for it, and one of the most successful racing bikes ever was a six. However it has a number of disadvantages, apart from its complexity and manufacturing costs. Its width and the length of its crankshaft are really the limiting factors; a tuned version could be expected to run into crank failure problems as they are almost traditional with this engine format.

12. V6
Not as mechanically perfect as the in-line six, the V6 motor is rather more practical, being more compact and having a much shorter crankshaft. It is probably the logical progression from four-cylinder layouts, depending on the balance of the power, cost and complexity equation.

13. V8
The V8 is probably the ultimate piston engine, putting maximum displacement and piston area into the most compact space and combining it with good balance and smooth power delivery. Its complexity and associated installation problems are its main disadvantages.

While some of these layouts may seem theoretical, they have all been used by motorcycle manufacturers since 1950. Other configurations – notably V10 and V12 – give still more piston and valve area at the expense of yet more length and more complexity. While engine balance is an important factor because it allows lighter castings and chassis components, firing intervals are also significant. Designers saw 180° twins and V-twins as bad, while 6-cylinder engines (and upwards) were good, because the former had uneven firing intervals while the latter could have lots of equally spaced pulses which should provide optimum traction and the smoothest drive.

In practice, whenever traction is a problem, this doesn't work. The equally spaced firing probably does give more traction but when the tyre breaks away the wheelspin is then reinforced by the regular arrival of firing pulses. The rider finds this hard to control. Conversely, having unequally spaced pulses may make the tyre break away a little earlier when two pulses arrive in quick succession but the following gap gives it the opportunity to recover and makes it easier for the rider to control. American flat trackers in the early 1970s found that 180° twins gave better drive than 360° versions of the same engine (with firing intervals of 180°–540° and 360°–360° respectively). In 1991–92, 500 cc GP riders found the same thing when their V4 two-strokes became too violent for current tyre technology; the V motors, especially those with two crankshafts, make it easy to get pretty well any firing interval at all, and when the intervals were made unequal – one short, one long, the so-called big bang engines – the riders found they got better, more controllable drive out of corners. The penalty was more severe loading in the driveline, with gear and primary drive failure becoming more common.

Evaluation

For a given engine, power production will be proportional to piston area, so one of the first things to consider is the possibility of increasing piston size. Big bore versions will show an immediate increase in torque and the larger bore also leaves room for bigger valves or for a wider squish band, which may be used to improve combustion or to allow higher compression before the engine suffers from knock.

The usual result of fitting bigger pistons is that peak power is moved further *down* the speed scale, so while torque is increased, peak power may not be. The reason is that the original valve time-area is no longer sufficient to deal with the increased displacement and, to maintain air flow and torque at high speeds, it will be necessary to find more valve time-area. Obviously the total area of the valves is of no consequence if they are not open, what is of value is the amount they are open combined with the time which they spend open during each cycle. This time-area concept is explained in the Appendix, along with a computer program to calculate it for given applications.

The next step is to compare the engine with other power units; first with engines in the same manufacturer's range, with a view to using parts from a larger or sportier version (similarly this may reveal that the engine *is* the larger version in which case it may already be close to its mechanical limits). The next comparison is with competition versions from the same manufacturer, to identify any similarities or compatible parts; then with state-of-the-art competition bikes, to establish a target.

The comparison should include piston area, valve area, valve time-area, cam duration, carburettor intake area, airbox intake area, airbox volume, peak speed and mean piston speed at maximum revs. Is there room for improvement in any of these factors? Does there seem to be a mismatch between any of them?

At some stage during this preliminary investigation it will be helpful to find out whether the engine has any weak points. As engines are developed further there is a tendency for parts to break; the most common being rod failure, piston failure, big end failure, valve failure, twisted or broken crankshafts and primary drive failures.

At this point it should be possible to formulate a target output. There is the choice of increasing the air flow – which means that any existing restrictions have first to be identified and then removed – or of sustaining the air flow at higher speeds, in which case it will be necessary to establish the safe maximum speed which the engine will accept.

An alternative approach is to prevent the air flow from falling off once the engine has reached peak speed – unless this phenomenon is going to be used to prevent the motor over-revving.

Bore size

The maximum size which the engine can accept may be governed by the displacement it gives; if not, the engine limitations revolve around the mass of the new pistons and the thickness of the liners. If new liners are to be fitted into a bored block then the spacing between the liners may become critical. Larger pistons will weigh more and will limit the safe maximum engine speed if the same rods and big ends are going to be used.

Finally there has to be adequate gasket area to make a seal which is oil tight or pressure tight, allowing room for any oilways which run through the block, or oil return passages, which sometimes use the stud holes.

Valves

The bore size and engine displacement will put a physical limit on the valve size and, at the same time, set a demand on the necessary valve area in order to achieve enough air flow at peak engine speed.

Cams

As the valve area is largely determined by the physical size and shape of the head, the engine has to achieve a suitable valve time-area by opening the valves for a long enough period. Time-area can be calculated using the program given in the Appendix; note that this is an experimental formula as far as four-strokes are concerned, a similar calculation predicts two-stroke performance very accurately, but this has not been generally applied to four-stroke engines. One difference between two-stroke and four-stroke valving is that two-stroke timing is symmetrical about the dead centre positions. In a four-stroke the same duration can be used for any number of valve open/closing positions. Naturally the timing of the valve is as important as its open duration and it has to be assumed that this will be optimized and will also be matched to other relevant factors which, put together, determine the engine's state of tune.

The valve time-area is determined by the cam profile and the cam follower, so once a figure has been settled, this dictates the form taken by the camshaft, and its timing.

Cylinder head

Work on the ports and on the exposed portions of the valves often gives big increases on air flow rigs, but the same increases are rarely produced in dyno tests. This suggests that the air flow tests do not simulate real conditions, possibly because they only relate to steady state conditions and possibly because they are too severe and tend to exaggerate small differences.

The head, in conjunction with the piston crown, determines the combustion chamber shape and is critical to engine performance. It is also critical from the point of view of mechanical reliability, the valve-to-valve and valve-to-piston clearances diminishing as the cam timing is opened up and the compression ratio is raised.

There are other practical factors to be considered, such as the change in dimensions if the head or block is machined in order to raise the compression ratio. Effectively, the head is lowered and this alters the dimensions of head steadies and external oil-feed pipes; it can also alter the cam timing in an OHC motor because it shortens the driving run of the cam chain or belt (see Appendix).

Carburettors
Can the carburettors be altered to match the new engine specification? Sometimes it is not possible – either physically or within the rules governing the class, and then the carburettor – or whatever proves to be the most restrictive component – becomes the main factor on which the rest of the engine specification is based.

Crankcase
An increase in engine displacement or speed will probably need more capacity in the crankcase breather – even in a four-cylinder engine in which there is no overall displacement, the movement of gas is enough to absorb energy and an efficient breather will sidestep some of this problem. Changes in the running clearance at the crankshaft bearings and the piston skirt may also create oil control problems and it may be necessary to use a different type of piston ring or to modify the lubrication system with baffles in the sump or, ultimately, to use a dry sump.

Reciprocating parts
Inertia forces generated in reciprocating parts are proportional to the mass of the part and the square of the engine speed. Not only do they cause vibration, they also absorb power because the forces needed to move the parts have to come from somewhere. In the same way, a rotating part requires a force proportional to its mass in order to accelerate it from one speed to another. And heavy parts naturally create greater frictional forces in the bearings which have to support them. Consequently if the weight of moving parts, particularly reciprocating parts, can be reduced, engine power will be saved.

Reliability
When an engine is uprated it puts three kinds of increased stress on its working parts. The fatigue loadings are increased, particularly those caused

by the inertia forces of the reciprocating parts – which generally increase as the square of the speed. Second, the drive loadings are increased. Third, thermal loads are increased, which has a direct bearing on some parts plus several side effects, for example, where temperatures run higher the working clearances are likely to change and, at high temperatures, the strength and fatigue strength of the materials are reduced.

Producing a realistic target

From these considerations it should be possible to produce a realistic target, the engine being built to match the weakest/most restrictive part. The relative difficulty of changing various parts should also suggest a likely development programme producing one-step-at-a-time changes for the minimum number of engine rebuilds. Where possible it will help to have two engines, or duplicates of major assemblies like cylinder heads, so that development can progress in a series of leap-frog steps keeping the option of reverting to the previous step when one modification goes too far and causes a loss of power or a component failure.

The order of power increase over a standard engine, the associated disadvantages and the chapter reference are shown in Table 3.1.

Table 3.1 Potential power changes

Change	Increase (%)	Problems	Reference Chapter
Blueprinting	0 to 5	availability of parts	10
Intake/air cleaner	0 to 15	more noise, less midrange	7
Exhaust	±10	more noise, less midrange	6
Cams	5 to 10	loss of power below peak	5
Big-bore kit	5 to 10	reduce peak speed; vibration; oil leaks	4
Raise c.r.	5	detonation	4,8
Carburettor	5	poor low-speed power	7
Full race development	30 to 50	illegal for road; loss of flexibility and reliability	–

It is worth noting that, generally, the losses will exceed the gains, the trick being to incur losses in areas which are not of use, or which can be tolerated.

Chapter 4

Pistons, liners, combustion chambers

In a seriously modified engine, the piston is the limiting factor. It plays a key part in determining each of the following characteristics:

1. Piston area.
2. The mass of the piston limits the safe maximum crank speed.
3. Piston and ring design put a limit on maximum piston speed.
4. Squish clearance, combustion chamber shape and compression ratio.
5. Thermal limits, particularly in the design of the crown and in the material used for the pistons.
6. Frictional losses and oil control.

Every aspect of power delivery is affected by one of these and so any changes made to the engine also have to be accommodated by the piston design. The ultimate stage of engine development will be closely followed by piston failure, or piston-related failures. Often it will come down to a matter of strength – usually the ability of the crown to withstand thermal loads – and eventually, more strength will mean more weight and this in itself will put a lower limit on the engine's operating speed. As there is always a strength/weight conflict, piston materials and design are critical.

Forged pistons are, traditionally, the strongest and, because of their greater density they tend to make better heat sinks and can transmit thermal loads more readily. For the same reason, they also tend to be heavier than cast pistons.

Cast pistons have two advantages. They can be light and their design is less restricted. It is easier to design a more complex construction where, e.g. extra metal may be needed in specific parts of the piston to provide added strength.

The best materials, in terms of strength/weight and anti-scuff properties, are high silicon alloys, but these are difficult to cast and machine. However, the Japanese have made an art of this and their standard cast pistons compare very favourably with special competition parts available from aftermarket firms. It is certainly worth using the stock pistons initially and to keep them until the engine reaches a state of tune which leads to piston failure. Even then there is no guarantee that a better replacement will be found. Typically the stock pistons will suffer from melted crowns; forged replacements may well prove tougher, but will also be heavier and will possibly transfer the problem from the pistons to the connecting rods or big-end bearings.

The only likely improvements are through the use of new materials. Already pistons are appearing with coated skirts and crowns while ceramic inlays are being used to toughen the pistons thermally. As the manufacturers are in the best position to make use of new materials technology, it is likely that the OEM pistons will still have an advantage over replacement types. It can also mean that pistons will be more difficult to modify. Some engine builders have pistons and other parts shot peened to toughen them, work-hardening the surface, and closing up small, stress-raising notches or cracks in order to improve the fatigue strength of the piston. If this is done to a coated piston, it will remove the coating and probably weaken the part.

Piston area

Increasing the bore size is an easy way to enlarge the engine, raising its capacity both as a pump and as a heat exchanger. It is limited by the physical thickness of the liners (or the distance between the liners) and by the engine's capacity class. There are several side effects, some good, some not so good.

The advantages are that the wider bore automatically gives a higher compression ratio and produces a wider squish band – something which many four-strokes lack. It also gives more room in which to fit bigger valves, especially on two-valve hemi head designs.

The disadvantages are that the new pistons will be heavier, placing a lower safe limit on the crank speed and creating more vibration. This may be countered by machining the flywheel webs or the balance shafts (the calculations for this are given in the Appendix). As far as performance is concerned, bigger pistons will reduce the specific time-area of the valves, so, while the increased air flow should give more torque, it is likely that the point at which power peaks will be shifted down the speed scale. Bigger valves or more cam duration will be needed to offset this and allow the engine to breathe efficiently at its original maximum speed.

If the conversion is made by boring the existing liners, they will be thinner and more susceptible to distortion and fatigue failure (see below).

Fig. 13. Components of a big-bore kit

Fig. 14. Additional methods of securing a shrink-fit liner in a cylinder block. Fatigue failures in liners usually occur just below the flange.

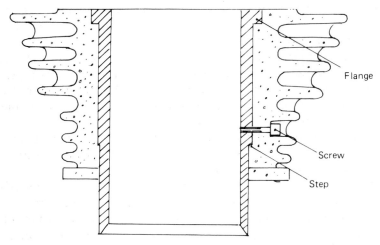

They will not be able to conduct heat away from the piston quite as efficiently and may well cause it to run at a higher temperature.

The alternative to this is to bore the block to take bigger liners, assuming there is room between each pair: flanged liners may have to have a flat machined on them to allow them to butt together.

The last disadvantage of a big-bore conversion is that new gaskets will have to be made up (see below) and there will be less gasket area. This often causes minor oil leaks, which are unsightly rather than likely to cause any damage.

The minimum thickness for iron liners can be gauged from the manufacturer's stock items; add the wear limit to their biggest oversize and subtract from the liner thickness. What's left is considered to be the minimum, plus a safety margin for road mileage. On a competition engine it would be possible to reduce this margin – but not too far, because it will eventually create thermal and/or fatigue problems. Depending on the bore size, the minimum liner thickness will be in the region of 3 to 4 mm, with an absolute minimum of 2.5 mm for engines which only need to run for very short periods.

Fatigue occurs because the liner is not such a tight fit in the block at the engine's working temperature and, under the force of gas expansion, the liner will tend to balloon outwards. If there is a flange at the top, this part will be much more rigid than the rest, yet the maximum force occurs some 15 or 20 mm down from the top, so any bending or shear forces set up are going to be concentrated into quite a short section of the liner. If the liner's strength has already been reduced by over-boring, then the cyclic flexing may go past its endurance limit and it will fatigue, cracking a short way down from the flange.

This could be avoided by making the liner thicker and (if the block is bored to accept bigger liners, by increasing the interference fit). In borderline cases the rest of the engine can be protected from this type of failure by positively locating the lower section of the liner, preventing it from falling down if it should break below its flange.

This can be done either by machining a small (0.4 mm) step in the lower part of the liner and a corresponding step in the block, or by pegging the liner. To do this, a hole has to be drilled through a substantial part of the block, near the bottom, and a thread tapped in it. A set screw is then screwed through into the liner, before boring.

When either the block or the cylinders are bored, it should be done using a machine which can keep all of the bores parallel and square to the axis of the crankshaft, rather than centring itself on the existing bore.

After the block has been bored and before the liners are fitted, it may help to have it bead blasted to peen the surface, in order to seal the inner surface and avoid the risk of oil stains seeping through and discolouring the outer surface.

After pressing the liners into place and checking that they are all at the same height, they should be bored to size if necessary, and finally honed to give the required piston clearance. On a tuned motor this would normally be slightly more than the production clearance – an increase of around 0.013 mm on a 50 mm bore being a typical figure.

If the type of piston has been changed, follow the manufacturers' recommendations for piston clearance. Pistons made in high-silicon alloy will have a low rate of expansion and can run tight piston clearances (as little as 0.03 mm). If the piston is made in low-silicon alloy it will have a high coefficient of expansion and so its cold clearance needs to be greater (up to 0.09 mm). The pistons should be matched for weight and individual sizing. Some manufacturers do not bother to get all the pistons to exactly the same size, in which case they must be measured individually and each cylinder then bored and honed to suit its piston.

Honing is a skilled process which does more than bring the cylinder up to its final size. It also leaves a cross-hatched pattern which forms the initial bearing surface for the piston and rings and which has to hold a film of oil. Bedding-in, subsequent wear and oil control are highly dependent on this finish. Hepworth and Grandage recommend a cross-hatch angle of 60° to the bore axis, with a plateau area of ½ to ⅓ for their cast-iron liners.

On assembly, check that all pistons reach maximum height at TDC; if not test the crankshaft for twisting or the cylinder block for truth. If the piston heights themselves vary, it may be necessary to machine the pistons to get equal values. As an alternative to linered bores, some engines have aluminium bores armoured either by chrome plating or by a process such as Nikasil. There are advantages in light weight, better thermal conductivity and a rate of expansion which matches that of the piston, but it does mean that the block cannot be rebored. Piston clearance is adjusted by selective assembly and it is important to use the type of ring recommended by the manufacturer. Mahle, in Germany, will recondition Nikasil bores as long as the base surface is not damaged or scored.

Piston mass

The maximum acceleration reached by the piston is when it goes through TDC. At this point, the acceleration A is:

$$A = b^2 r \left(1 + \frac{r}{L}\right)$$

where b = rotational speed in radian/second
r = stroke/2
L = rod length
(b = $N \times 0.1047$ where N is the crank speed in rev/min)
(see Appendix)

The force F generated at the big-end and the lower part of the connecting rod is given by

$$F = mA$$

where m = the mass of the piston, rings, piston pin, circlips and the top two-thirds of the connecting rod.

The manufacturer will, hopefully, have determined the largest force which the connecting rod can withstand and this, for a given piston mass will fix the maximum safe operating speed, which is where the manufacturer puts the red line on the rev counter.

If the piston mass is then changed, the maximum safe speed will also have to be changed to achieve the same level of stress. For the same stress in the rod,

$$b_2 = b_1 \sqrt{(m_1/m_2)}$$

where b_2 = new max speed
b_1 = old max speed
m_2 = new mass of piston, pin, etc
m_1 = old mass of piston, pin, etc.

Piston speed

The mean piston speed V is

$$V = 2SN$$

where S = stroke
N = crank speed

As a rule of thumb, a general maximum mean piston speed, without trick piston rings, is 20 m/s. At this speed, conditions such as oil control and the amount of friction at the cylinder wall, change. Friction and oil drag become significant and it may be worth switching to a two-ring piston – which will not seal or give as good oil control at low speeds.

The other problem associated with high piston speeds is ring flutter in which the inertia of the ring is great enough to lift it off the bottom edge of the groove as the piston reverses direction. This causes sealing/oil control problems and rapid wear. Various types of ring and their characteristics are shown in Fig. 15.

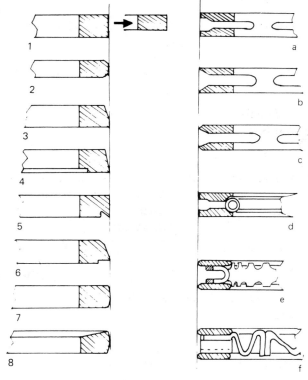

Fig. 15. Some types of piston ring, showing designs intended to reduce ring flutter or to give better bottom-edge contact for more oil control, or to allow rapid bedding in.

Compression rings: 1. Plain – most frequently used top ring, going to thinner sections to avoid flutter at high engine speed. 2. Plated, often with bevelled face, gives good wear but cannot be used on plated bores. 3. Taper faced – increases bottom edge contact and rapid bedding-in. The taper is only about 1 degree. 4. Internal steps and bevels make the ring twist slightly when fitted to the bore, either to put a higher load on the bottom edge or, as in this case, to allow gas pressure to act on the top ring face, forcing the ring into harder contact with the cylinder wall and the bottom edge of the groove. A tapered or barrelled face is necessary to permit the ring to twist. 5. Externally stepped (Napier). The shape and twist in this design usually give better oil control. 6. Externally stepped and tapered. A progression on (5), both usually used as second rings. 7. Barrel faced. A ring prone to flutter or rapid wear takes on a shape similar to this, so the as-new shape follows the wear contours and should prevent scuffing. 8. Internal taper, barrel faced. A design intended to allow the ring to withstand flutter in high-speed engines, without losing contact with the cylinder wall and causing blow-by. As the ring is lifted off the bottom edge of the groove, it tends to twist, forcing the lower edge outwards into firmer contact with the wall. The barrelled shape has to permit twisting without losing wall contact.

Oil scrapers: (a) One piece grooved ring. (b) heavy duty – the smaller lands give a higher wall pressure. (c) narrow land – a further progression from (b) giving approximately 250 per cent more wall pressure than (a) and 150 more than (b). (d) Helical spring-backed, grooved ring. A means of increasing the wall pressure without using very narrow lands. May be 60 per cent higher than (c). (e) Apex ring. This is one of Hepolite's most popular oil-control rings, using separate, thin steel rails, separated by an expander which increases the wall pressure as it is compressed. (f) Hepolite SE. Similar to the Apex ring, but with independent expander units which do not allow the ring to be compressed vertically.

Combustion chamber

This needs to achieve two objectives, to give as much compression as possible without causing knock and to give good combustion. Now 'good' means as fast as possible, but in an orderly manner, and with the minimum of heat loss.

This sets the first requirement – a sphere has the minimum surface area for a given volume, so the basic shape of the combustion chamber should be spherical. It is immediately compromised by the piston, whose crown is the weak point and here the ideal shape is flat, or just slightly curved (up or down). The result is the hemi head, for a long time the standard combustion chamber shape, with widely angled valves to get the biggest area – until the valve seats touched one another in the centre and ran into the edge of the head at the sides. In this layout the spark plug has to be offset (ideally, combustion should begin at the centre of the sphere).

In order to get high compression the piston then has to have a fairly large dome and the dome needs to have cutaways in it so that the big valves will not touch it on the overlap stroke. This all weakens the piston crown, so it has to be made much heavier and the combustion chamber becomes the three-dimensional equivalent of a long, thin crescent, which gives the flame a long path to travel. The surface area has also increased, so that heat loss is greater and thermal efficiency smaller and the piston has lots of edges around the cutaways – edges which run hotter than the rest of the piston and cause pre-ignition.

A modified version of the hemi-head consists of two part-spheres, the larger one housing the intake valve (see Fig. 17). This is arranged purely to get a bigger valve on the intake side without wasting volume on the exhaust side.

In fact the wasted volume to each side can be filled in, creating a squish band where the piston clearance at TDC tends to zero.

Squish turbulence is very useful. The effect is made by creating a highly-compressed ring of gas around the edge of the cylinder, which forces the outer regions of gas to move rapidly in towards the centre. This gas is then moving towards the combustion flame while its own turbulence makes it more willing to burn. The result is that combustion is rapid and more complete than it might otherwise have been. The evidence for this is that the specific fuel consumption will improve, while the ignition timing can be retarded without a loss of torque.

A secondary effect is that the temperature of the combustion chamber may be reduced, which will, in turn, permit a higher compression ratio to be used before the engine suffers from knock. If these benefits are not seen, then the squish turbulence is not working.

It seems that piston-to-head clearance is critical if the squish band is to work. Ideally the clearance should be just above zero, which means that when the engine is stationary and cold there will be a clearance of 0.6 to

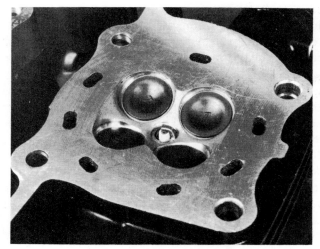

Fig. 16. This Honda head has the valves recessed, giving high compression and presenting a fence between intake and exhaust. The shape of the outlet from the valve also affects gas flow through the valve and swirl inside the cylinder

1.0 mm, depending on the engine construction and materials. During the development stages it is often possible to reduce the clearance progressively until there are signs that the pistons are just touching the head at maximum crank speed and to go back to the previous stage; gas turbulence also prevents deposits from building up on the head, another sign that the squish band is having some effect.

Measuring the static clearance is not always easy; the usual methods are to use modelling clay or a strip of soft solder trapped between the head and the piston as the engine is rotated and to measure the depth of squashed material. The results should always be checked by examining the pistons and head for signs of contact after the engine has been run.

Four-valve heads offer a more compact combustion chamber, with less surface area, stronger piston crown design and more opportunity to use squish turbulence. The early types used on production motorcycles (in the late 1970s) did not reap the full benefits because they still featured a large included angle between the intake and exhaust valves. This affects the shape of the combustion chamber in the head and also demands deep cutaways in the piston crowns to give enough valve clearance.

Setting the valves at a shallow angle gives the compact, pent-roof design which is a more practical alternative to the ideal, spherical shape. Having two smaller intake valves can mean that more valve area is obtained within the same bore size, but this is not the main advantage. Maximum valve lift (see Chapter 5) is proportional to valve diameter, so less lift is required to reach full air flow. Piston-to-valve and valve-to-valve clearance become less critical; also, because the valves are lighter, they can be lifted faster without over-loading the valve gear. This gives a large benefit in valve time-area plus the ability to run at higher engine speeds without the risk of valve, or valve-train, failure.

There are two more advantages. First the spark plug can be sited in the centre of the combustion space (in fact it would be difficult to put it anywhere else). Second, the area occupied by the valves represents a square, roughly, within the circular boundary of the head. the D-shaped segments left at the sides can make an excellent squish band.

Heads can be modified by reprofiling or by filling with weld and then profiling, while it is often necessary to recut the valve seats or to sink the valve further into the head. The easiest method of raising the compression ratio is to mill the gasket surface of either the head or the cylinder block.

Having modified the head in one of these ways, it will then be necessary to check its volume, for future reference or to ensure that all of the combustion chambers are equal. The head volume alone is enough for this sort of comparison, but if it is necessary to know the compression ratio, then the volume of the piston crown must also be checked.

First the head volume can be measured using a burette with thin liquid in it. The head must either be level or a flat, transparent plate must be bolted to it, with a hole in the centre large enough to let in fluid and let out air. If the head is set level it is only necessary to fill it until the fluid surface is level with the gasket face. However there can be an error here, caused by surface tension in the fluid forming a meniscus which makes it difficult to judge when the two surfaces are level.

To find the volume of the piston crown it is necessary to measure how close the outer edge of the piston crown is to the top of the block at TDC (sometimes called the deck height). This, in effect tells how far the piston protrudes into the combustion chamber. Then, with the top ring fitted, set the piston a short way (h) down the bore, so that all of the crown is within the cylinder, and seal the gap between the top ring land and the barrel with grease. Now the cylinder can be filled from the burette and its volume measured (V). The volume of a plain cylinder of the same height would be $\pi d^2 h/4$ where d is the diameter of the cylinder. Consequently the volume of the piston crown above the top land is $\pi d^2 h/4 - V = V_p$.

If the deck height below the top of the block is a and the thickness of the head gasket (when squashed) is g, then the combustion chamber volume will be

$$V_h + V_a + V_g - V_p$$

where V_h is the measured volume of the head
          V_a is the calculated volume of the cylinder above the deck height, that is $V_a = \pi d^2 a/4$
          V_g is the volume caused by the head gasket, that is $V_g = \pi d^2 g/4$
          V_p is the volume of the piston crown (see above).

If this total combustion chamber volume is called V_c then the compression ratio will be $(D + V_c)/V_c$, where D is the piston displacement for that cylinder.

The type of head gasket will also need to be considered at this stage. There are several choices, ranging from the stock item through made-up sheet copper gaskets to special gaskets.

The original gasket is the easiest to use, but it may be inadequate or cannot match the new bore size. Plain copper gaskets sometimes have an advantage in conducting heat away from the head and may help if the engine has a problem with pre-ignition or detonation. Various other types of gasket are available from performance shops, including wedge-shaped or pressure-backed versions and gas-filled rings. Some of these types need a locating groove to be machined around the edge of the bore.

Copper gaskets should be annealed. The traditional method is to put a smear of soap on the metal and then heat the gasket uniformly until the soap turns brown. With all non-standard gaskets, make sure that there is also a seal where any coolant or oil passages pass from the cylinder block to the head. If a thicker compression gasket is used, it may mean that these other gaskets will not be compressed properly and an alternative, such as a thicker O-ring should be used. Use a high temperature sealant such as Hylomar on the head gasket surfaces.

Thermal limits

To avoid damage, the piston material either needs to be able to withstand thermal shock (see below), or it must be able to conduct excess heat away, and cope with distortion due to uneven expansion.

Its ability to dissipate heat depends on its contact with the cylinder wall, via the rings and the lubricant. Consequently the type and number of rings is important (see Fig. 15) and so is the wall pressure of the ring. Some oil-scraper rings are available in a variety of wall pressures, while other small variations can be made by using a slightly over or under-size ring, re-gapped to suit the new bore. Using two rings instead of three will tend to make the piston run at a higher temperature.

Skirt clearance depends on the piston's original size and on its expansion relative to the bore. Because the piston is neither heated nor constructed uniformly, it tends to distort out of shape when the engine reaches its operating temperature. The idea is to produce a piston which will match the shape of the bore when it is hot, and so the cold dimensions tend to be irregular. Usually pistons are tapered, being wider at the skirt, because the crown will expand more. They are also cammed, having a larger diameter at right angles to the piston pin. For information on measuring pistons and piston clearances, see Chapter 10.

Pistons with slotted skirts should not be used in high-performance engines. Instead, changes in skirt clearance should be accommodated by honing the cylinder barrel and any oil control problems tackled by changing the oil-control ring, drilling oil relief holes in the skirt or by modifying the lubrication system.

Fig. 17. Comparison of head designs.
Top left: two-valve hemi. The plug cannot be centrally located when big valves are used and a large piston dome (with deep cutaways to avoid the valves) is needed to get a high compression ratio. The combustion chamber is very long and thin.
Top right: Forming the head on two radii allows the intake valve to be made bigger, set at a smaller angle, and gives a more compact combustion space.
Bottom left: the four-valve, pent-roof design is more compact still, has a central spark plug and, because the valves are smaller and do not need to be lifted so far, does not need deep cutaways in the piston crown. The shape also allows a broad squish band to be formed (shaded area).
Bottom right: A five-valve layout, as used by Yamaha, gives a compact chamber with part-spherical surfaces, reducing the surface area for a given volume. The squish area is greater than the four-valve design

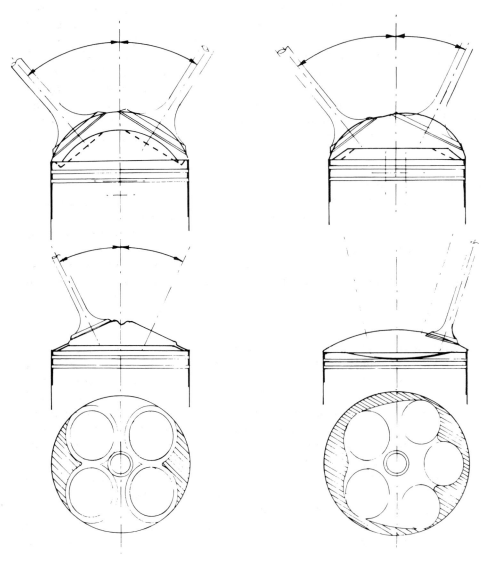

Raising the thermal loads on the piston may mean that its original shape is no longer quite correct. Increasing the skirt clearance will not solve the problem if the crown is expanding too much – in fact, it can make things worse, because it will allow more piston rock, forcing the ring lands to bear against the cylinder wall. If the ring lands show signs of scuffing or smearing, it may be necessary to re-proportion the piston. A small amount can be machined from the ring lands, extending down to just below the oil-control ring. The amount should be as small as possible – in the order of 0.013 mm – although the exact amount can only be found by trial and error.

Overheating problems, such as piston holing, melted crowns or melted top lands can sometimes be solved by better piston cooling, either through the use of different rings or by improving the supply of oil to the underside of the piston. Most engines have a splash feed from the big-end journals, and this will be increased if the journal is run at a greater radial or axial clearance. Some engines also have an oil spray, from jets in the oil gallery feeding the crankshaft bearings. This supply could be increased slightly by enlarging the jet size. The alternative is to use a stronger piston or reduce the heat flow to it, by retarding the ignition, running a richer mixture, lowering the compression ratio, or lowering the back-pressure in the exhaust system.

Frictional losses

Lubrication and oil control have already been mentioned, but there are a variety of piston designs which are meant to reduce friction, usually by cutting away part of the skirt (slipper piston), shortening the skirt, or putting raised pads to take the thrust loads on the skirt. These measures are usually a compromise because they make for worse oil control, do not allow heat to be transferred to the cylinder wall so easily or allow more piston rock.

New materials

Ceramics and sintered metals are likely to play an increasingly large part in the design of engine thermal components. Already the low thermal expansion and scuff resistance of silicon is exploited in piston alloys, while sintered piston rings have allowed a greater degree of performance and reliability. Ceramics may have a direct influence on spark-ignition engines, although it is more likely to be an indirect effect stemming from the work done on turbines and diesel engines, both of which suffer higher temperatures and greater thermal loading.

Ceramics, the traditional materials of the pottery industry, were initially developed by engineers for their refractory properties and from this stemmed their use as insulators and dielectrics.

Their main features are: high strength, although they are brittle and will not bend or be deformed; high resistance to corrosion (especially compared to metals at high temperatures); high resistance to wear; lower coefficient of expansion (which means that it is possible to run smaller clearances); and they are light (about one-third the density of steel). Their brittle nature is obviously one disadvantage, although even this can be put to some use, for example, in the case of high-speed turbine blades. If there is a failure due to impact or to a flawed component, turbine blades tend to do a lot of damage. Rotors made from ceramics, however, break up into small particles and therefore do not need an armour-plated housing.

The major attraction, though, is their strength at high temperatures, which means that critical parts of the engine need not be subject to the limitations of the cooling system. In fact they may not need to be cooled at all. This, in turn, might raise the low thermal efficiency associated with piston engines.

Ceramics can be split into two groups, oxides and non-oxides. Oxides include silicates, such as aluminium silicate, while the non-oxides are made up of nitrides, carbides, borides, etc., formed on a base material such as silicon.

The manufacturing process is to mix the material in powder form and then fire it at very high temperature, or melt it, in the case of things like glass. The problems are largely in joining different parts and in ensuring uniform quality throughout the material (as brittle fractures are sudden, any slight defect could lead to immediate failure in use). As the manufacturing process has a lot in common with the sintered metal processes, it is possible that one will benefit from the other as techniques are developed. The most likely components to use ceramics are:

- turbocharger rotors and entry pipes
- cam followers
- valve guides
- valve heads and seats
- pistons

While the initial attraction is in the material's high resistance to thermal shock, its lightness and good wear properties also make it attractive for highly stressed reciprocating parts, such as valve gear.

Chapter 5

Valves and cams

Valve area provides the basis for air flow, and the amount of valve open duration dictates the engine's power characteristics. Unfortunately, as far as the calculation is concerned, the valves are not opened instantaneously; a considerable part of the open period is spent in the act of either opening or closing the valves. As the presence of the valve in the port is a continual obstruction to gas flow, its shape and the shape of the port around it can make a significant difference to the flow, particularly during the part-open stages.

During most of the valve's open period, flow is directly proportional to lift. When the lift is increased beyond a certain point, the flow no longer increases but settles at a constant level. At this point the head of the valve is no longer the most restrictive part in the gas flow.

Depending upon the proportions of the head of the valve, its stem and the discharge coefficient of the port, the effective area in the throat of the valve will be equal to the effective area of the port when the valve lift is between 0.26 and 0.31 times the diameter of the valve head. Also, the valve seat region forms a nozzle, discouraging turbulent flow. Above a certain lift this will no longer apply and turbulence may occur.

Tests on air flow rigs confirm this, usually indicating a maximum effective lift in the region of 0.26D to 0.28D for large valves. Smaller and better streamlined valves may benefit from slightly more lift, in the order of 0.30D or more, but a flow rig is really necessary to indicate whether any changes in port or valve shape would warrant an increase in lift.

Any further lift is of no benefit to the air flow and only puts extra stress on the cam and cam follower, but, just as the valve gear cannot open the valve instantaneously, nor can it suddenly hold the valve at a particular height. In practice it is necessary to reach the optimum valve lift as quickly as possible without over-stressing the cam follower, and then to let the valve decelerate, so that maximum lift will always be greater than optimum lift.

Optimum lift, plus the tolerable acceleration of the valve train (see Appendix) will determine the basic dimensions of the cam and its follower. But power is not only proportional to valve area; the valve opens a certain area for a certain amount of time. It could achieve the same gas flow by having less area and opening for a longer period, or by having more area and opening for less time. If a graph of valve open area were plotted against time, the area under the graph would represent the time-area integral and this

Fig. 18(a). Proportions of the valve head and seat. At low lift, the valve area forms a nozzle (A) which expands until – at optimum lift – it has no further effect. If the seat is cut back (B) the valve can be sunk into the head. The new profile can give more valve-piston clearance and improve gas flow.

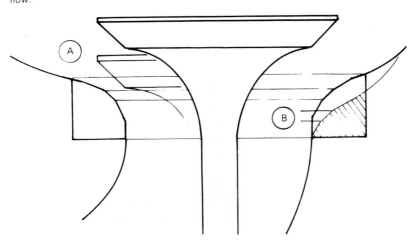

Fig. 18(b). Air flow tests on a 500 cc Honda four-valve head, with a pressure drop of 10 in of water across the intake valves. A – standard (1.374 in) valve and port: flow reaches a maximum at 0.35 in lift, giving a L/D ratio of 0.255. B – standard valve, modified port: flows more air after the valve has lifted 0.15 in, but still reaches a maximum at 0.35 in lift. C – larger (1.406 in) valve in the modified port; the flow is increased, and now reaches a maximum at 0.4 in lift, giving an L/D ratio of 0.28

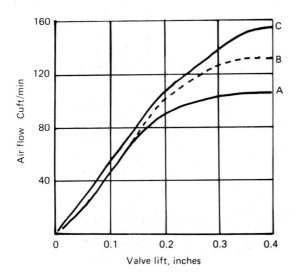

Fig. 19(a). There is a variety of valve shapes which give different flow characteristics; the best flow depends on the shape of valve in conjunction with the shape of the port and the exit into the combustion chamber.

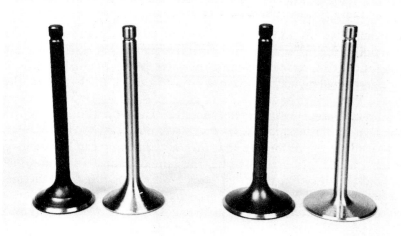

Fig. 19(b). Yamaha's five-valve heads (*left*, FZ750; *right*, FZR1000) combined the maximum intake valve area with the best proportions for combustion.

would dictate how much air could flow through the valve under any given conditions.

Because the time available for each phase is inversely proportional to crank speed, the time-area diminishes as the speed increases; at 8,000 rev/min there will be half the time-area available at 4,000 rev/min.

As the valve spends so much time in the part-open state, the calculation for time-area is tedious. However, the BASIC program in the Appendix can calculate it from valve lift and valve size data. If the valve is lifted beyond the optimum point, the program treats it as if it were fixed at this point. It is possible to get more time-area by opening the valve earlier or closing it later, or by opening the valve faster, but not by opening it further than the optimum.

This program is something of an experiment as far as four-stroke engines are concerned. A similar program works well when applied to two-strokes, but four-strokes have the difference that the valve opening can be timed to the engine (or 'phased'). In a two-stroke, the port opening is symmetrical about TDC and the duration therefore fixes both the opening and closing points. In a four-stroke this does not apply, and there is plenty of evidence that the actual timing points have a strong influence over the power characteristics. Figs. 20 and 21 show graphs of intake duration versus peak torque speed, and intake closing point versus peak speed. Obviously this relationship has to be kept in mind when altering the time-area figures.

The time-area can be used in several ways. First, to find the time-area at the engine's peak torque (where the air flow per cycle will be at a maximum) and then to see what changes would have to be made to get the same time-area at the target speed for peak torque in the modified engine.

Second, the figures can be used to compare one engine with another, even allowing for different capacities, as the program gives specific time-area — that is the time area divided by the piston displacement.

Third, the specific time-area figure can be used to see how the cam or valves might have to be changed if the engine size is altered. In this way it can predict an engine's state of tune — but it has to assume that the rest of the engine matches the valve timing and that the opening and closing points are acceptable to the engine.

In order to fill the cylinder at high engine speeds it becomes necessary to open the valve faster, or to open it earlier. There is a limit to how quickly the valve train can accelerate the valve — valve acceleration is also calculated by the program in the Appendix — and as this is proportional to the square of the engine speed, it is not difficult to reach a speed at which the valve gear will become unreliable. If it is not possible to open the valve faster, then it has to be opened earlier and, to match high engine speeds it is common to find extremely long valve open periods.

There is an obvious disadvantage, that, at low speeds the valves will be open for so long that intake gas will mix with the exhaust or that gas will be

Fig. 20. The empirical relationship between intake duration and engine speed: (*left*) the speed at which peak power is produced; (*right*) the speed at which peak torque is produced, for a variety of engine sizes

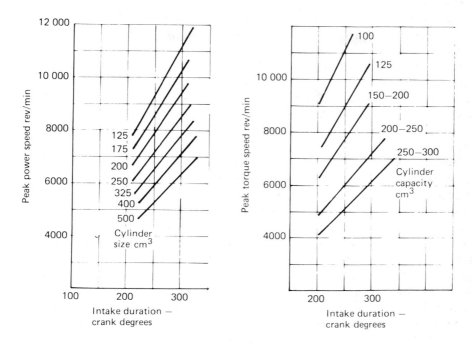

Fig. 21. The empirical relationship between the intake closing point and the speed at which peak power is produced

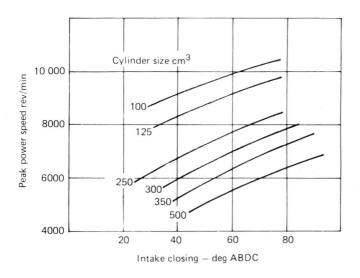

blown back past the intake valve instead of being trapped inside the cylinder. Low speed performance falls and eventually the engine will become difficult to run.

However, long valve duration can also have an advantage. At some speeds the cylinder filling and engine performance will increase out of proportion to the time-area figures. The speed at which this happens is governed by the length of the intake system and by the exhaust system. The pulses set up by the valve opening and closing are presumably reflected back and forth along the ports and the returning pulse will, at some speed, fall into step with the generated pulse. One will strengthen the other and the overall air flow will increase.

This is quite useful, but it only happens within a limited speed range and, to either side of this speed range, the air flow and power are reduced considerably by comparison. This sudden step in the power is not particularly desirable if the object is a smooth, flexible, easy-to-control engine. It can be avoided by making the pipe lengths so short or so long that they never resonate with the engine in its normal speed range. Or the ports/pipes can incorporate steps and sudden section changes which disrupt the pulses and dissipate the energy in them. To make use of the pulses, the ports must be kept smooth and it will be necessary to experiment with the overall length of the port/pipe system — both for the intake and the exhaust. The benefit is a sizeable power increase. The penalty is an equally abrupt drop in power at speeds either side of the resonant period.

Stock Japanese engines generally have a high output combined with a long rev range and very flexible power characteristics. They do this by using valves which are light and with cam profiles which open the valves very rapidly. The time-area required is achieved by a combination of valve-area and rapid valve opening. The duration and valve overlap is very conservative; often there are devices to disrupt pulses in both the intake and exhaust.

It is possible to take advantage of this, modifying the intake and exhaust to work at resonance and using cam profiles with a longer duration. Peak power is increased but the power band is narrowed considerably.

The shape of the port and the valve affect gas flow and while it is possible to describe optimum shapes in general terms, detail changes can only be investigated by experiment. Cylinder head specialists use an air flow rig (see Chapter 2) and often make a mock-up of the port in wood.

The first requirement is that the port should be smooth and straight with the minimum of obstruction. Where there is an obstruction, e.g. throttle plates, valve guides, it should be streamlined; that is, its width, as viewed by the oncoming gas, should be small in relation to its length. Tapering the ends of valve guides usually has no measurable effect; streamlining the boss holding the guide might be worthwhile.

The absolute size of the port is a matter for compromise (apart from physical limitations) because while a larger port can flow more gas, a smaller port will increase the gas velocity; the kinetic energy of the gas and the corresponding changes in pressure can be quite useful. As a rough guide, the optimum carburettor size approximates to $d^2/40 = p$, where d is the bore of the carb in mm, and p is the horsepower the cylinder produces; CV carburettors need to be about 20 per cent bigger in diameter. Smaller bore sizes give better midrange power and generally make the carburettor easier to tune.

This size then dictates the proportions of the intake port. The gas flow from the carburettor hits the valve stem and is splayed out over the valve head into the cylinder. To avoid turbulence in the critical region leading up to the valve head, the port is usually tapered gently to make it form a shallow nozzle, usually around 80 per cent of the maximum cross-section area.

The shape of the valve head can obviously have a major effect on air flow, particularly during part-lift, and as the valves spend more time on part-lift than they do at full lift, this can make big differences to the engine's performance. There are several basic shapes available, ranging from the almost flat, 'penny on a stick' shape, through to the so-called 'tulip' type. The shape which will give most flow depends on the shape of the intake tract, the angle at which it emerges into the combustion chamber, and possibly on the proximity of neighbouring valves.

If the valves are being tested in an air flow rig, it is possible to build up the profile of the head using modelling clay in order to find the most suitable shape.

When Yamaha developed their FZ engine, they examined configurations ranging from 4-valve up to 7-valve heads, before settling on the 5-valve design. Keeping all the valves within the boundary of the cylinder bore, and allowing a minimum of 2 mm between intake valves and 4 mm between exhaust valves, they maintained a ratio of 1.3 between the intake and exhaust valve area. Within these constraints, the 5-valve head gave 20 to 24% more intake area than a 4-valve head, and it also gave more intake area than the 6- and 7-valve head designs which were evaluated.

They built 4- and 5-valve engines with the same bore and stroke, and also a bigger bore 4-valve which had slightly more valve area than the 5-valve, but to accommodate this, the bore size became wider than what they considered to be optimum for the combustion chamber shape. The other engines had a bore/stroke ratio of 1.16, while the big-bore, short-stroke version had 1.59. (When Yamaha went into production with the 1985 FZ750, it had a bore/stroke ratio of 1.32 but by 1990 the FZR750 was up to 1.57 which they kept for the 1993 YZF750. In the meantime the Suzuki GSX-R750 had a ratio of 1.63 in 1988, which they reduced to 1.44 in 1990. Between 1989 and 1993 the Kawasaki ZXR750 had crept up from 1.32 to 1.50. Both the FZR and the ZXR motors needed work on combustion – higher compression, tighter squish clearance, ignition timing optimized – before they would respond to further tuning. This implies that a bore size of 71–72 mm with a bore/stroke ratio of 1.5 or a shade more is, as Yamaha said, the maximum width for a combustion chamber. There are additional considerations, mainly that shorter strokes and longer connecting rods create less inertia forces and reduce engine friction – see Chapter 1.)

Yamaha's engine tests showed that the 5-valve had better volumetric efficiency than the 4-valve with the same bore, and the combustion was as good in both engines. Compared to the big-bore 4-valve, the 5-valve had roughly the same volumetric efficiency and it had better combustion. Throughout its rev range, the 5-valve had better BMEP (or torque) than either of the 4-valve engines. Finally, because the individual valves were lighter, the 5-valve could be run some 700 rpm faster than the 4-valve before it suffered valve bounce or float.

One curious anomaly emerged from Yamaha's air flow tests. Their figures show that the four-valve engines reached maximum air flow when the valve lift/valve diameter (L/D) ratio was in the region of 0.29 to 0.31, which agrees with general experience. The 5-valve head, with its smaller valves which are also closer together, reached an L/D ratio of 0.37 to 0.38 at maximum air flow. It is possible that this was caused by interference between the air streams from neighbouring valves, in which case, changing the shape of the valve head and the port exit, to angle the flow downwards could improve the flow still further.

On any engine the region of the valve to the valve seat and the entry into the head, can promote turbulence or swirl in the gas. There is conflicting evidence here, some suggesting that swirl is beneficial, and some that it is not. It is likely that some swirl will help the gas clear the valve seat area, avoiding throttling and perhaps guide the intake gas away from the still-open exhaust. Swirl in the wrong direction could also take gas out through the exhaust. Turbulence in the throat of the valve would clearly spoil the flow. Swirl also helps to distribute the fuel mixture more evenly and to speed up combustion. However, swirl continuing after combustion will also bring more hot gas into contact with the metal of the engine, increasing heat loss and reducing thermal efficiency. It seems likely that a certain amount of swirl does improve performance; more swirl will then cause a reduction in efficiency.

The lengths of the intake and exhaust will have to be matched to the speed range of the engine; many formulae have been produced, but always with the proviso that they only give a starting point from which to begin experiments. They might as well say make the intake to the carb bellmouth 30 cm (12 in) long and make the exhaust headers 81 cm (32 in) long and start from there.

Whatever effect the exhaust and intake pulses have, will depend ultimately on the opening and closing points of the valves. The ideal cycle, bearing in mind that suitable conditions only exist within a narrow speed range, is:

The intake opens while the exhaust is still flowing. Good discharge characteristics of the exhaust port and low pressure in the exhaust system will keep the exhaust gas moving in the right direction and prevent any tendency for it to change and wander out of the intake valve. A strong, high-pressure pulse in the intake will start its gas flowing and, as long as the gas does not flow straight from the intake into the exhaust valve, the incoming gas will push out the remaining exhaust gas. At this point the piston is still rising. Low pressure in the exhaust will draw the gas out of the cylinder; the sudden arrival of a high-pressure pulse will stop the flow, at which point the exhaust valve needs to close.

Intake gas continues to flow, its velocity/swirl enabling it to carry on flowing into the cylinder after the piston has reached BDC. A high pressure pulse in the intake port will help and the valve should close just as flow comes to a standstill, trapping the maximum amount of gas inside the cylinder.

After combustion, once the piston has picked up speed, the gas expansion has a diminishing effect. If the exhaust valve opens suddenly, the gas will still have enough energy to burst out into the port; the optimum point to open the valve is when the gases' energy is of more value as an exhaust pulse than as a means of propelling the piston.

As the theory suggests, the best results from valves and their timing come from painstaking experiments, each coming a little closer to the optimum combination. The practical aspects are:

Ports

Where there is a curve in the port it should be gradually tapered into a nozzle, to avoid turbulence; the tighter the curve, the more taper it needs. Any steps, bumps or changes of section should be ground out if pulse effects are going to be used; they should be built in if the pulses are to be destroyed.

If the intake valves close early, gas will still be travelling in the port and a surge tank can be used to some advantage. This can take the form of a pipe downstream of the carburettor leading into a chamber, or leading into another intake port (sometimes called cross-porting). A pressure transducer in the port would show how the pulses bounce back and forth and, while the valve is closed, some of the oscillating pressure energy can be used. A surge tank will simply soak up the pressure pulses, preventing them from doing any harm. Cross-connecting the ports can transfer the energy in No. 1 intake to No. 2 intake. The theory suggests that this could work at any speed or throttle opening; practical results show little difference at wide throttle openings – perhaps because the cross connections cannot flow

Fig. 22. Optimum intake port proportions: straight and smooth, tapering to about 80% of the maximum area at the point where the restriction to flow is greatest.

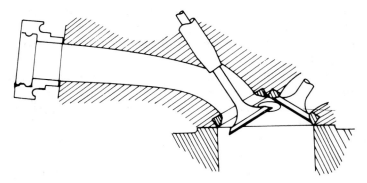

enough gas to make any real difference. However, they can make a large difference at small throttle openings, increasing the power while reducing the fuel flow – either because the size of the cross-port is large in relation to the throttle opening, or because it makes the pulses in the carburettor less severe and by evening out the flow in the carburettors it improves both the

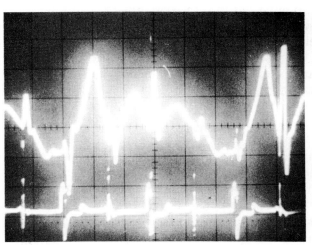

Fig. 23. *Top*: oscilloscope trace produced by a pressure-sensitive transducer in the intake port of a 250 cc cylinder running at peak air flow. The lower trace is from the ignition, to synchronize the top trace. When the radio-frequency 'spikes' have been taken out, the pressure trace looks like the graph below, which clearly shows the pressure fluctuations after the valve has closed, caused by the pulse travelling back and forth along the port

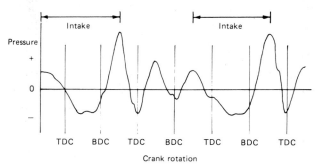

fuel delivery and the pick-up. This could be a useful way of reducing the unwanted effects of pulse tuning.

Generally it is all too easy to grind away metal in the wrong places. Bifurcated ports and valve guide bosses should be made as *long* as possible in the direction of the gas flow, while the bore of the port should be closely matched to the stub, gasket or pipe which joins it.

The entrance to the intake port should be made of a fully radiussed trumpet, even if this is enclosed within the air box or inside an air filter. It provides a smooth entry to the carburettor and a definite end to the intake tract, a point from which pressure pulses will be reflected. If the wall of the air box or filter is very close to the carb entry, it may have a bad effect on either the gas flow or the carburation. This sort of problem can usually be cured by spacing the air box further away from the carburettor. Other problems, such as fuel spraying from the bellmouth, can be cured by changing the length of the tract.

Valves and guides

An engine's valve time-area can be increased by altering the cam profile or by increasing the size of the valves. A two-valve engine will be restricted by its head size ultimately, while one in which the valves have a large included angle will also be restricted by valve-to-valve and valve-to-piston clearance. For a given valve area and lift, the air flow can possibly be increased by streamlining the valve and its seat.

Valve lift and the clearance around the valve head can be measured by using a dial gauge and a degree disc fitted to the end of the crankshaft. The gauge must be mounted securely, with its stem parallel to the stem of the valve and its follower sitting on the top spring collar. The most common methods are to make up an angled bracket which bolts to the cam box studs, or to bolt a flat steel plate to these studs and use a magnetic stand to hold the gauge.

First locate TDC using a dial gauge to trace the movement of the piston by finding the point at which the dial gauge reverses its travel and zero the degree disc to this point. From now on, only turn the engine forward, particularly if the camshaft is driven by chain or by any mechanism which may have backlash.

Most cams have opening and closing ramps, that is portions between the base circle and the lobe of the cam which open or close the valve gently. In these regions the rate of lift is very low and it is difficult to get reliable measurements for lift or for opening and closing points. Consequently it is common practice (a) to increase the tappet clearance to a wide setting, typically 1.1 to 1.3 mm (so that the entire ramp is used up in closing the valve clearance) or (b) to measure the open point at 1 mm of lift (which achieves the same end and involves less engine building if the valve clearance is adjusted by shims).

Having established the valve's opening point relative to one of the dead centre positions, turn the crank another 10 degrees and measure the lift here. (Any angular interval can be used – the program in the Appendix needs lift data every 10 degrees). Continue until the valve is deemed to have closed; it may involve turning the engine several times to find the position at which the valve has returned to 1 mm of lift. When making these measurements it often helps to have only the inner valve spring fitted and, if the engine uses a generator with a permanent magnet, to remove the rotor. You will also need to know the method used by the manufacturer or the cam supplier for timing the cams. Both of the above methods are used for establishing the open point; many cam specialists prefer to 'phase' the cams, that is to establish the point at which the valve reaches maximum lift. Finally, it is necessary to check that all valves on all cylinders are following the same lift pattern. It has been known for cam grinders to get slightly out of phase between one lobe and the next.

The same process is used to check the clearances between the valve and its neighbouring valves and the piston. To do this a small lever is used, either to pry between the cam and the follower or to push down on the valve/rocker arm. In the region of 40 degrees either side of TDC at the top of the exhaust stroke and the beginning of the intake stroke, push the valve further open until it touches something or until the spring goes coilbound. If new cams have been fitted it will also be necessary to check that the main spring does not bind at full lift; it should have at least another 1.1 to 1.5 mm of travel. The spring's initial pre-load and its fully compressed length can be adjusted by either machining or shimming the bottom spring seat.

Valve-to-piston clearance should be more generous at the exhaust valve, as it will be closing and may float, i.e. lose contact with the cam, at high speeds. The minimum acceptable clearance will depend on the size of the valve and the size of the piston and the connecting rod. Bigger engines stretch further. A generally accepted minimum is 1.5 to 2.0 mm, although some racing engines run smaller clearances. For a given cam profile, extra piston clearance can be obtained either by sinking the valve seat back into the head or by machining cutaways in the piston crown.

Valve-to-valve clearance depends largely on which valve is in front. The exhaust valve is closing and, if it floats it will naturally be further away from its seat than the measured distance. The intake valve is being opened and so will follow the cam motion. Consequently if the exhaust valve is in front of the intake, the clearance should be quite generous – in the order of 2.0 mm. If the intake is moving across in front of the exhaust it may be possible to let the gap close up slightly, say to 1.5 mm. Valve-to-valve clearance can be increased by sinking the valve seats or by altering the phasing of the cams, if they are separate.

The fitted height of the valve and spring can become an important measurement. What it amounts to is the height of the collet groove above the bottom spring seat and it can be altered in several ways:

1. By fitting a longer or shorter valve
2. By machining the bottom spring seat or adding steel shims to it.
3. By cutting back the valve seat.

The valve itself can sometimes be shortened by grinding the end of the stem, as long as the stem remains proud of the valve collar and the grinder does not break through the layer of hard metal.

This fitted length affects the spring seat pressure and how close the spring gets to coilbound when the valve is lifted (see below). It may also affect the geometry of the rocker arm; the further over-centre the rocker goes, the less effective it is.

It is only logical that the shape of the valve head should influence gas flow and can therefore dictate the optimum amount of valve lift. Some tuners machine the exposed portion of the valve stem to reduce its diameter and also radius the valve head; it always seems like one of those logical steps which do not work in practice, or one that shows a difference on an air flow rig without making any noticeable change in power. Then Yamaha produced their five-valve FZ750, one of the best four-stroke designs available, as far as power is concerned. It has very straight ports and waisted valve stems.

The conclusion is that the valve stem will make a noticeable difference when it is the restrictive part of the system; if some other component is creating a greater restriction, or if the flow is turbulent, then this sort of modification will go unnoticed.

Valve clearance is necessary to ensure that the valve seats fully during its closed period. This is essential to conduct heat away from the valve head into the cylinder head, otherwise the valve will overheat and either wear rapidly or cause pre-ignition. When the open duration of the valve is increased, the closed duration is naturally decreased, reducing the opportunity for valve cooling. As there will be more heat energy floating about in a tuned engine anyway, this often leads to overheating problems. This, and spring overheating (causing the spring to sag and lose its tension) are common in turbocharged engines.

There are several possible solutions.

1. Increase the valve's contact with its seat, by increasing the closed duration.
2. Increase seat contact by making the seat wider (although a narrow seat is better for air flow – see below).
3. Improve the cooling in the region of the valve. On their turbocharged XN85, Suzuki opened an oil gallery below the exhaust valves and fitted ducts to the head to increase air cooling. Running a richer fuel mixture tends to lower the internal temperatures because the latent heat of the fuel uses up some of the engine's heat. Retarding the ignition and lowering the compression ratio also let the engine run cooler, although

Fig. 24. Modified valve guide (*top*) shown against the stock item

the last three items are really final resorts, aimed at side-stepping the problem rather than curing it.

4. Changing the valve and seat material for something tougher – which assumes that something is available – either from a race kit or adapted from another engine. Specialist suppliers' catalogues may help in tracking down the right parts.

The clearance used will depend on the operating conditions – the important thing being that some clearance is maintained under all operating conditions. Usually the stock clearance or a fraction more will be sufficient. Greater clearances will only make the engine noisier and will cause damage by the hammering effect they produce.

It is worth thinking this through clearly when any top end alterations are planned. Typical modifications will affect the valve clearance as well as the valve-to-piston clearances; for example, grinding the cam's base circle will tend to increase the clearance, possibly beyond the range of adjustment provided. This may make it necessary to sink the valve into the head, because this will also increase the valve-to-piston clearance which will diminish if the compression ratio is raised.

If the valves are changed, it may also be necessary to alter the valve guides in order to use a compatible material. If an austenitic steel (such as Nimonic 80) is used then the best guide material is bronze. For valves made in EN8 or a similar steel, cast-iron guides are best. The valve-to-guide clearance is difficult to measure and many manufacturers now rely on the 'rock' or 'wobble' method, in which the wobble is measured at the head of the valve. The appendix includes a calculator program to convert wobble figures into valve/guide clearance.

Finally, the collet groove in the valve stem is a highly stressed area and will itself raise the local stress. The groove must be fully radiussed, with no sharp, internal corners.

Valve seats

The seats are usually cut at 45 degrees but there are exceptions. The face width will usually be around 1mm although exhaust valves may need slightly more if there is evidence of burning or sinking (sinking or 'pocketing' can also be caused by valve float, in which case stronger springs should be used).

The seat width is adjusted by first cutting the seat in the head, using a 45-degree cutter to get an even face or to sink the seat to the required position. A single-point cutter is the best tool to use if the seat is to be sunk an appreciable distance.

Wide and narrow angle cutters (usually 15 and 60 degrees) are then used to trim the seat back to the required width and to centralize the seating face on the head of the valve. The valve is then lapped on to the seat using grinding paste, and the seat area and evenness of contact should be checked with marker dye.

Valve springs

The weakest possible springs should be used, short of letting the valve float. Usually the stock springs will be adequate.

When the valve is being opened, it will reach a peak velocity, depending on the cam profile and the engine speed and, shortly after, the nose of the cam will take the load off the follower, leaving the valve spring to slow down the valve and to make it follow the receding profile of the cam. At this point, the force trying to make the valve continue its motion is the inertia of the valve, collet, collar, follower, and the top half of the spring. The momentum of these parts is countered by the spring force and what happens next is a matter of simple arithmetic.

The contact pressure between cam and follower will drop enormously, possibly to zero or less (i.e. valve float). The spring must now bring the valve to rest and return it, ideally keeping light contact with the cam. Wear marks on the cam often suggest that the valve 'takes off' at the nose and lands (fairly heavily) at the closing ramp. If it lands beyond the ramp it will soon damage its seat – an instance where a stronger spring would actually protect the valve and seat. The BASIC program in the Appendix calculates valve train velocities and can match spring forces to the inertia loads in the valve train.

Cams

Cams can be changed by one of the following methods.

1. Phasing – or the timing of the intake and exhaust relative to one another. Even a single cam engine can have the whole shaft advanced or retarded. Basically, retarding the intake cam will tend to improve high-speed running at the expense of low-speed running. Advancing the cam timing will give more low-speed torque and reduce high-speed torque. In essence, the engine will give the same shape of torque curve, but the effect of altering the cam timing will be to tip it slightly, as Fig. 25 shows. If the manufacturer has got it right in the first place, then the losses will be greater than the gains, as the diagram also shows.

To set the cam timing (sometimes called 'degreeing-in') it may be necessary to use slotted sprockets or some other means of varying the relationship between cam and drive gear (see below). It will also be necessary to have a valve lift diagram or precise information on either the opening point or the point at which the valve reaches full lift (see above). Check whether these figures are obtained with an increased tappet

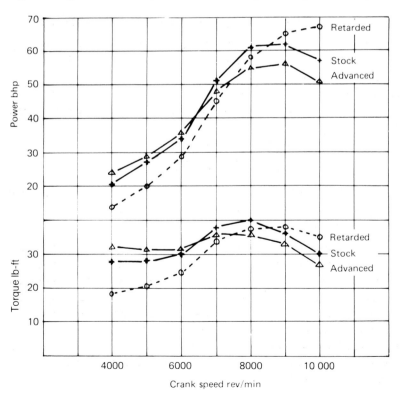

Fig. 25. The general effects on performance caused by advancing and retarding the valve timing

Fig. 26. Vernier sprocket mounting for the cam drive

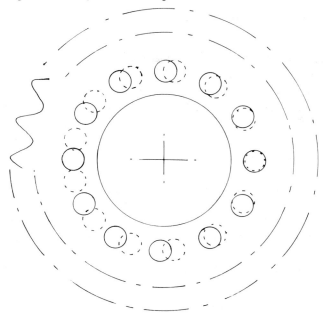

clearance or if they assume an opening point at, say 1 mm lift. Next:
(a) check that the cam chain or belt has been correctly tensioned, or that any gears/shaft have been shimmed to give the correct backlash. Only turn the engine forwards.
(b) Use a degree disc and dial gauge to locate TDC – on the correct cylinder and stroke! – and set the engine in the approximately correct timing position.
(c) Move crank forward to the exact position and hold.
(d) Move cam until valve is in required position.
(e) Tighten cam mounting bolts.
(f) Turn engine through two revolutions and re-check. Check other valves.
(g) Reset tappet clearance as necessary.
(h) Repeat for second camshaft.

There are several ways of holding the drive sprocket to the cam shaft and some which can be modified to allow the kind of adjustment mentioned above. Simplest are the slotted sprockets, which are usually available as performance accessories or could be made up by a machine shop. The sprocket should be checked for eccentricity and should ideally be located by a shoulder or spigot, so that the bolts are only used to clamp it.

Sprockets located by a peg can be modified by making a vernier coupling (see Fig. 26) which can give adjustment in 2 or 3 degree increments, an offset peg can be made, or a new hole can be drilled.

Fig. 27. Offset pegs and keys can be used to alter the timing slightly

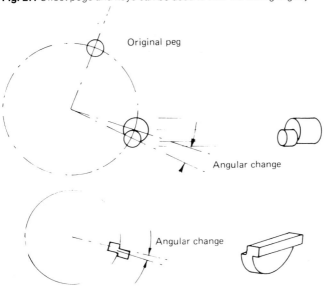

Where sprockets are keyed to the shaft, an offset key can be used.

If the sprocket is mounted on a taper, then it can be adjusted to any position, although this type of fixing is rare on motorcycles.

2. *Regrinding*: in this process it is only possible to remove metal (i.e. from the base circle) which immediately gives more lift and more duration, although this can be controlled to some extent by the new profile (see Fig. 28). It will be necessary to take up the increased clearance in some way, either through the valve adjuster (which may mean having special shims made up) or by altering the fitted height of the valve.

If the cam is cast iron ('chilled' iron) then only the nose of the cam will be hardened and the base circle can be ground easily, with no need for further treatment.

Steel cams, on the other hand, are case hardened, with an outer skin which is around 1 mm thick and so they will need to be re-hardened by heat treatment after any grinding.

3. *Re-profiling*: in this process the cam is built up by weld or metal spray and then ground down to the new profile. It has the advantage that there is practically no limit on the new profile and it need not effect the base circle dimension. The disadvantages are that it is more expensive and the cam will need heat treatment to harden it afterwards. The process can also cause distortion, so it will be necessary to check the shaft for truth using V-blocks and a dial gauge, and re-straighten it in a press if necessary.

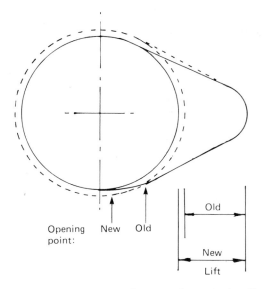

Fig. 28. Grinding the base circle of a cam will increase its lift and, if the same profile is followed (top edge) the duration will be increased slightly. If a new profile is followed (bottom edge) it can be increased dramatically

4. *New cams*: either made specifically for the model or bought as solid billets and machined to a particular profile. A larger cam lobe can run off the edge of the cam follower, and this must be avoided by increasing the size of the follower, which will give more rapid valve opening, or by reducing the radius of the follower, although this in turn reduces the rate of cam lift (see Figure 29).

Cam follower

The shape of the follower is critical as it affects the rate of lift at the valve, as well as the contact stress between itself and the cam. It can be used to increase time-area by opening the valve faster; basically the larger the radius of the follower, the faster the valve will be opened.

Where a cylindrical or bucket follower is used, the cam is often tapered so that it will make the bucket rotate to even out wear. In this case the face of the follower must be machined to a spherical radius of at least 1,250 mm.

Kits are available to convert bucket and shim followers to the more reliable shim-under-the-bucket type.

Where rockers are used, the arms may be of unequal length, giving the rocker ratio X : Y shown in Fig. 30. This may be as high as 1.4 (that is, the valve lift could be 1.4 × the cam lift) and about 2.0 in single-sided, or 'finger', rockers. The ratio can be altered to some extent by boring the rocker bearing off centre.

The angular motion of the rocker must also be taken into account if a longer/shorter valve is being used, or if the cam height is changed appreciably. The rocker transmits motion less effectively when its arm makes an angle of less than 90 degrees to the valve, etc. Changes to the geometry of the valve gear may mean that the radius on the rocker's working surface must be altered.

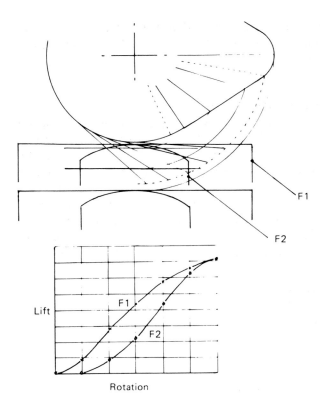

Fig. 29. The shape of the cam follower affects the valve lift as shown

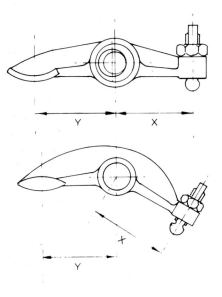

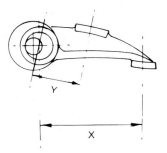

Fig. 30. The proportions of the rocker arm also affect valve lift; the rocker ratio X:Y gives the motion of the valve in relation to the motion of the cam. Boring the rocker bearing off centre (*top*) can alter the ratio slightly

It is often possible to lighten rocker arms and, at the same time improve their fatigue strength by polishing or peening the surface after removing any stress raisers (see Appendix).

Valve train velocity/acceleration/wear

Because inertia stress goes up in proportion to the square of the engine speed and as the valve components will be accelerated harder by the use of high-performance cams, several factors can become critical. It may be necessary to lighten the valve gear to prevent failure or valve float and suitable items for this treatment are: valve collars (lighter material), valve springs (fit with the close-coiled end away from the collet) and rocker arms. Parts should be stress relieved (see Appendix) to avoid fatigue failures.

There can also be problems with excessive contact stress at the cam follower or rocker arm, or at the valve seat. Any of the following steps will increase the contact stress:

(a) new cam form giving greater acceleration
(b) using higher crankshaft speed
(c) reducing nose radius of cam
(d) reducing radius of cam follower
(e) reducing valve seat width
(f) using stronger valve springs
(g) using heavier reciprocating components
(h) using a follower radius which is incompatible with cam taper
(i) increasing the rocker ratio

Failure or abnormal wear may be attributed to one of these factors and can be alleviated by altering one or more of them, or by increasing the surface contact area concerned.

Also, changes in dimensions may affect the lubrication system by obscuring or re-directing an oil spray, particularly where oil is sprayed from rockers on to the lobes of cams or to valve stems.

Cam forms which are too violent or are mismatched with the valve springs will allow hammering between the valve and valve lifter and possibly between the valve and its seat, causing rapid wear or failure, breaking the valve or the top collar/collet arrangement. If this is due to the valve not following the cam motion then either the cam profile or the valve spring will have to be revised.

Spring surge can also permit the valve to lose contact with the cam. This is caused by resonance – by the operating frequency hitting the natural frequency of the spring or, more probably, one of its harmonics. A stiffer spring, with a higher frequency, or an inner spring, will cure this problem.

The Appendix gives calculations for valve-train velocities and accelerations, along with a BASIC program which calculates them, and shows how this can be used to calculate the ideal valve-spring rate or predict conditions in which float will occur.

Cam drive

The vast majority of motorcycle engines have overhead cam shafts driven by chain – either a thin roller or a HyVo chain. There are a few alternative arrangements, gear drive kits, shaft or belt drive but these are in a minority.

The problem with chain drives is not the load but the speed. Some drives, including the light HyVo used on the 600 Kawasaki, run to speeds of 12,000 rev/min quite happily; others seem to have trouble operating at their standard design speeds. The basic rule seems to be: the lighter the chain the better it is suited to high speeds. The chain tensioner also needs careful attention. On competition engines it is common practice to change automatic tensioners for the variety which can be locked up manually, to avoid the risk of the ratchet mechanism backing off during use.

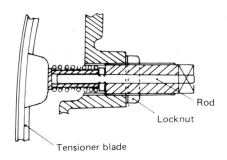

Fig. 31. A manual cam chain tensioner, as used on the early Z250 Kawasaki. It is set so that the end of the rod is flush with the adjusting screw

Lowering the cylinder head, e.g. to raise the compression ratio, will also cause the cam timing to be retarded if it is driven by belt or chain. A program in the Appendix gives the calculation for the change in timing. If the drive is by shaft or gears, some means will have to be found to accommodate the dimensional change.

Chapter 6
Exhaust system

Gas flow through the engine is subject to pulses, sometimes fairly severe pulses. There are effects, some of them useful, caused by the energy in the gas and by the pressure pulses travelling through it.

A pulse travelling through gas in a plain pipe will be reflected when it reaches the end of the pipe; if it is a solid end, the pulse simply bounces off it and proceeds in the opposite direction. If it is an open end the pulse pops out of it and a reflected pulse goes back along the pipe; the difference is that a low-pressure pulse will be returned as a high-pressure reflection, while a high-pressure pulse will produce a low-pressure reflection.

If the plain pipe is connected to a tapering pipe – a diffuser or a megaphone – it seems to have a similar effect to an open pipe. If the angle of the diffuser is very large, then it behaves just like an open pipe. But if it is narrow – up to an included angle of about 15 degrees – it seems to extend the length or the duration of the pulse.

The engine generates fairly strong pulses each time the exhaust valve opens. On a tuned engine in which the exhaust opens both rapidly and early, the pulse will be stronger still. The same pulses will be reflected back along the exhaust and the engine will be sensitive to the returning pulses if (a) the valve is just opening and gas flow is low (so a drop in pressure downstream of the valve will help and will strengthen the following pulse); (b) the intake valve opens while the exhaust is still open. In this case low pressure in the exhaust will encourage the burnt gas to continue on its way and not to mix with the incoming fresh gas. As the intake flow starts, high pressure in the exhaust, just before the valve closes, will prevent the new gas escaping from the cylinder. If the pressure pulses can arrive at the right time to help with any of these processes, volumetric efficiency will be increased. Obviously if they arrive at the wrong time they can prevent exhaust gas escaping or encourage fresh gas to mix with the exhaust; volumetric efficiency will decrease.

The time taken for the pulses to travel through the exhaust depends upon their relative velocity within the gas and the length of the exhaust system. These values are more or less fixed, except that the speed of the gas itself will vary with engine speed.

The time interval between the pulse being generated and the *need* for a reflected pulse to arrive will be inversely proportional to engine speed, so that chances of the two coinciding are limited to one speed or, in practice, to a narrow range of speeds. The effect can be seen quite clearly if an engine is run on plain, open pipes and their length varied. At resonant speed the power will be increased dramatically. If tapered pipes are used, the effect is less dramatic but it extends over a wider speed range, as if the pulse were being stretched and, in doing so was losing its intensity.

If the rest of the engine is arranged to give peak efficiency at one speed, exhaust effects can be impressive but usually this is not desirable because it contrasts too sharply with the loss of efficiency at speeds to either side of the resonant speed. In practice it is better to have a slight mismatch and spread the beneficial effect over a wider speed range – in other words, the exhaust can also be used to tailor the power characteristics of the engine.

The next step is to discover that the exhausts of multi-cylinder engines can be inter-active; when 2-, 3-, 4- and 6-cylinder exhausts are coupled into a collector, there is a better power increase than if the single exhausts are optimized. A 4-1 system is best for peak power, while a 4-2-1 system gives a better spread of power. This immediately brings in the dimensions of the secondary pipe, which is simply another variable to play with.

Of course, the pulses will only be able to work if they are allowed to travel freely along the exhaust system. They can be dissipated by putting in steps, sudden section changes or chambers; the same pulses make noise and they can be reduced significantly by the same mechanisms employed in a silencer. Yamaha did just this when they put their EXUP valves into the exhaust systems of the FZR1000 and the OW01. Either a guillotine or a butterfly valve was used to block off part of the secondary pipe, just downstream from the 4–1 collector. It was operated by a servo motor and an engine speed sensor, so that the valve was withdrawn above a certain speed (about 6,000 rpm) and projected into the exhaust below this speed. The exhaust was tuned like a race system, to give beneficial effects between 9,000 and 11,000 rpm and this would normally create bad effects at 6,000 to 7,000 rpm when the pulses got out of phase. Instead, the EXUP valve acted like a baffle and prevented the pulses from being reflected back along the header pipes. The result was that the engine behaved as if it had a neutral exhaust through the midrange and a race exhaust at high speeds.

In practice, the exhaust is developed on a dynamometer or at the track, from a starting point which is either copied from an existing system or made up. There are various formulae, some theoretical, some empirical, usually based on something like:

$$t = 2LV/(V^2 - x^2)$$

where t = time interval between the pulse being generated and the reflection returning
L = length of exhaust pipe
V = velocity of pressure wave in (still) exhaust gas
x = velocity of exhaust gas in pipe

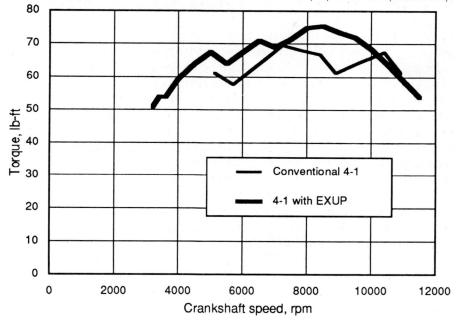

Fig. 32. Torque curves for a Yamaha FZR1000 with its original exhaust, including EXUP valve, compared to a conventional 4-1 system. The 4-1 exhaust, developed for peak torque at 10,500 rpm (and at just over 7,000 rpm), creates bad effects around 9,000 and just below 6,000 rpm. Had it been developed to give peak torque at, say 9,000 rpm then the inefficient region could be expected at around 7,000 – shortening the power band and possibly spoiling the engine's top end output. The standard exhaust is developed for peak torque close to 9,000 and although the torque curve shows small dips and peaks where the exhaust pulses go in and out of phase with the engine, the EXUP valve makes these fluctuations much smaller. The result is more torque plus a wider spread of torque.

the only snag is assumptions have to be made about the values of V and x.

If a guess has to be made about two of the variables, why not simplify things and guess L? In practice, the header pipes will be in the range 63 to 84 cm (25 to 33 in) – shorter for engines running at higher speeds – and the secondary pipe will be about half the length of the primary (a longer system will not conveniently fit a motorcycle chassis).

The practical aspects of exhaust development are as follows.

Exhaust port

For the exhaust system to be effective, there should be no dam, no change of section and the exhaust valve timing should have fairly long duration. The current practice among the Japanese is to fit their roadsters with a lot of valve area and cams which open the valves quickly, but which have fairly short duration. The result is a very long speed range and good top end power combined with flexibility. Sensitivity to exhaust systems is reduced and is often minimized by using section changes to dissipate pulse energy.

Fig. 33. A 4-1 exhaust system usually gives the best peak power, but not the best spread of power

Fig. 34. Collectors for a 4-1 system (*left*) and a 4-2-1 system (*right*)

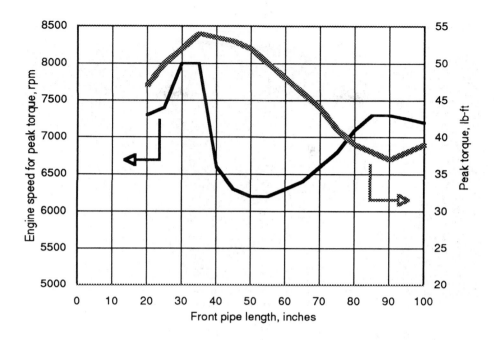

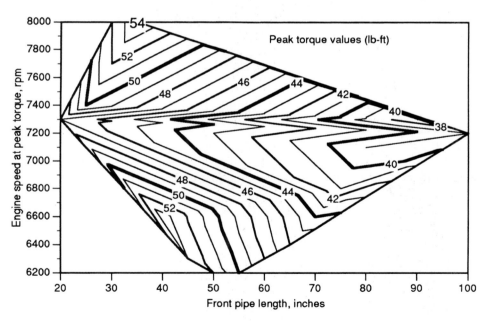

Fig. 35. Three ways of looking at the same thing: the effect of progressively lengthening the front pipe on a Rotax single cylinder engine – on peak torque and on the engine speed at which it made peak torque. The best length for maximum torque and power was about 89 cm (35 in) yet a length of 127 cm (50 in) only gave slightly less torque but brought peak speed down and gave the most flexibility. (The CB250/400 Honda twins had long rev ranges, with flat torque curves, using very long header

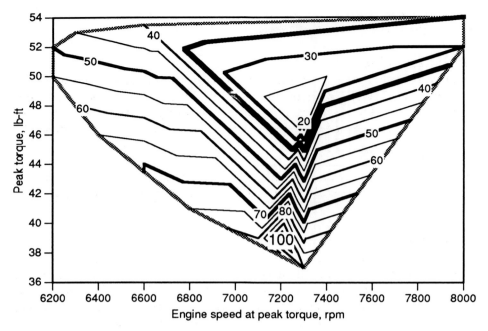

pipes. Their 2-2 exhausts had a large junction box joining both pipes just before the silencer cans; inside the box, the header pipes crossed over to the opposite side before finishing with a plain end, increasing the length of the pipe by almost 25 cm (10 in) and making the total length of the header - measured from the valve - somewhere between 101 and 127 cm (40–50 in). The 3-dimensional maps show (b) engine speed for peak torque v. front pipe length, the contour lines showing maximum torque values in lb-ft, and (c) peak torque values v. engine speed while this time the contour lines show the header pipe lengths in inches.

Header pipes

The most important factor is length, which can only be established by experiment. The diameter of the pipe is also important because the gas speed and lack of turbulence can be used to create low pressure near the port when the valve closes.

Obviously a pipe which is too small will be physically restrictive while a pipe which is too large will give low gas velocity and may cause turbulence. As the OE system usually works quite well, it makes a reasonable starting point; if a 10 per cent increase in torque is envisaged this will represent a 10 per cent increase in air flow and, to maintain the same exhaust properties, this will mean an increase in diameter of 4.8 per cent – or about 2 mm on a 38 mm pipe.

Collector

For exhaust pulses to work, the collector needs a smooth entry and exit. A ridge or a dam across the pipes here will allow them to behave as four separate pipes. Some collectors have a long axial baffle, which effectively makes the system a 4-2-1.

Secondary pipe
A large diameter is needed, but not too large as this tends to reduce the system's sensitivity to length. In any case, the effect is of a smaller order than that caused by the primary pipes. The secondary may run straight into the silencer, be the first element of the silencer or be a tapered diffuser.

Diffuser
Stretches the exhaust effect over a wider speed range, but does not give as much peak power as a plain pipe. It seems to be most effective on single exhausts.

Cross pipes
Where separate exhausts are used (e.g. 2-2, 4-2, 6-2, etc) a cross pipe will open the volume of both silencers to all of the cylinders. The effect of the stub pipe itself may attenuate some frequencies. There also tend to be increases in low-speed load and in the engine's pick-up and response. Cross pipes have been positioned close to the head or at a distance of some 76 cm (30 in) from the head; some engines have two cross pipes.

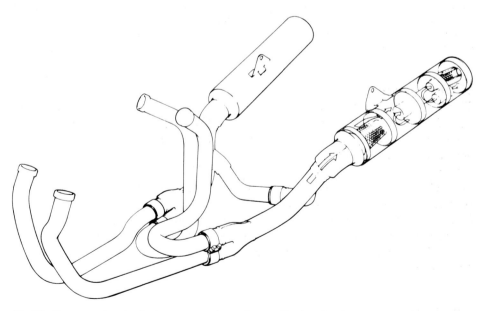

Fig. 36. Component parts of a Japanese roadster silencer. The lengths of each chamber or tube can be altered to attenuate a particular noise frequency, making the silencer effective without being restrictive to gas flow

Fig. 37. Two designs of silencers for getting the greatest effect from the smallest volume.
Top: a reverse flow silencer, patented by Leon Moss and Nick Stephenson, in which the gas enters a primary chamber and follows a tube to a secondary chamber, where the flow is reversed along concentric tubes, each one perforated. The size of the chambers and tubes can be altered to attenuate particular frequencies.
Below: a silencer designed by Dr Geoff Roe, in which the gas flow is kept to the outer passage. The inner compartments (Helmholtz chambers) are selected to take out particular noise frequencies and there is no physical obstruction to the gas

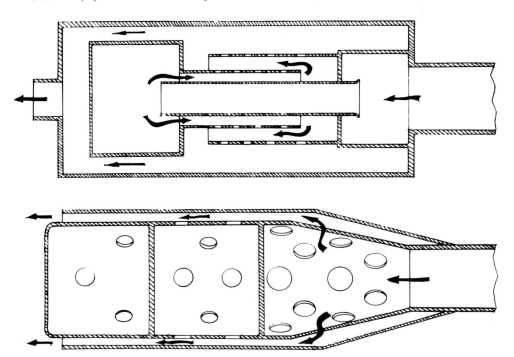

Silencer

Absorption type: a perforated tube leading into a can packed with sound-deadening material. This type has low back pressure but is noisy because it is not selective and simply takes some energy out of all frequencies. It is cheap to make and suitable for competition use.

Capacitive type: heavier, more expensive but much more efficient. This type contains a series of chambers interconnected by short tubes, each reacting to a particular frequency; it is efficient because it only needs to concentrate on the frequencies produced by the engine. Various designs use reverse flow to make the most use of the available silencer volume; some use active chambers, others use Helmholtz chambers. Where a small silencer can has to be used, the back pressure may rise significantly — enough to cause the combustion temperature to increase.

Finally, as well as altering engine performance, the exhaust can be used to improve all-round performance by saving weight and by improving ground clearance.

Chapter 7
Fuel system

The statement that power is proportional to air flow relies on the assumption that the optimum amount of fuel can be introduced to the air and be successfully burnt. Vague descriptions like 'optimum' and 'successfully' are necessary because these qualities vary with the operating conditions.

Optimum means an air-to-fuel ratio of about 12:1 by weight when full power is required, dropping to about 18:1 for maximum economy. The engine also needs a very rich mixture for starting and idling, with extra richness to cope with transient conditions like sudden acceleration. Actually the engine does not *need* the rich mixture but in these borderline conditions it is the only way to ensure that something like the right mixture arrives inside the cylinder. All of the air will get there but some of the fuel will be lost, dropping out of the air flow and forming a liquid layer on the sides of the intake tract. In steady conditions, this would soon evaporate and the rate of evaporation would equal the rate of drop-out; but in transient conditions such as starting or accelerating there is not the time available to reach a steady state.

In much the same way, the fuel should ideally be presented in a form which makes it possible to get complete combustion – that is, the fuel must be broken into tiny droplets ('atomized') and evenly distributed throughout the air. Otherwise the fuel will not be burnt in the short time available between ignition and TDC – typically around 0.0005 sec at high speed. The fact that this process is not totally efficient is demonstrated by the need to use an air/fuel ratio of 12:1, while the chemical balance for total combustion would be around 14:1.

From this it is possible to see that there are three mixture conditions, other than perfect. These are rich (less than 12:1), weak (in excess of 14:1), and 'wet' – in which the fuel is not sufficiently atomized for full combustion. Definitions of richness, etc, can be deceptive. For example in a rich mixture you would expect unburnt fuel to be found in the exhaust, while a weak mixture would leave oxygen in the exhaust; unfortunately, a wet mixture will give both of these symptoms, regardless of the proportions of fuel and air flowing from the carburettor. Consequently it is necessary to measure the fuel flow and relate it to the power produced; this result is the fuel flow (in lb/h or pt/h) divided by the power (in bhp) and is known as the specific fuel consumption (SFC) and is measured in units of lb/hp-h or pt/hp-h. The equivalent in ISO units is gm/kW-h.

This is a kind of measure of efficiency – what you pay for divided by what you get – and therefore the smaller the number the better. Experience shows that the optimum value is around 0.50 to 0.55 for most four-strokes. Below this the mixture strength is weak and the gas temperature goes up, locally – enough to overheat valves or piston crowns. At a value of around 0.3 to 0.4 pt/hp-h the mixture will be so weak that the engine will misfire; similarly, a rich mixture will cause misfiring in the region of 0.80 pt/hp-h. The actual values will depend on the efficiency of the carburettor as an atomizer and of the combustion chamber design.

Fuel flow measurement is usually made by putting a flow meter in the supply to the carburettor(s), in which case it is important to check that the measured fuel is reaching the engine and is not being lost through carburettor flooding or through being sprayed back from the carburettor bellmouth.

The SFC figure makes a useful comparator during engine development, as it can often point out restrictions in the engine. If, for example, the engine speed is increased and both the fuel flow and dynamometer load go up, then the air flow must have increased and all is well. If the speed and fuel flow increase but the load does not, then it suggests a restriction in air flow, upstream of the carburettor. If the speed goes up but the load and fuel flow do not, then the restriction is downstream of the carburettor.

The process for finding the ideal mixture strength involves running the engine at constant speed and throttle, and monitoring the load while the fuel flow is progressively decreased (by changing the appropriate jet, needle, etc.) The resulting chart of fuel flow v. load or power is called a mixture loop and from it the optimum setting for power or economy can be seen.

Table 7.1 Mixture loop at constant speed and constant throttle

main jet	fuel flow (pt/h)	power (bhp)	SFC (pt/hp-h)	
largest	49.5	66	0.750	rich misfire
	43.0	68	0.635	
	36.6	69	0.530	optimum
	30.1	68	0.443	
	22.6	66	0.342	
smallest	18.3	63	0.290	weak misfire

A slightly rich mixture is the easiest to ignite; the fuel burns quickly initially because the drops of fuel are packed more densely and the voltage requirement at the spark plug is lower than for a stoichiometric or even maximum power mixture. Of course, towards the end of combustion it would be less efficient because there would be an excess of fuel and not enough oxygen. So the best all-round mixture would be one that was slightly rich near the spark plug and had excess air at the furthest reaches. This is called a stratified charge. When the plug fires, the rich mixture ignites quickly, heating and compressing the gas around it and propelling unburnt fuel towards the weaker mixture. This region, despite squish turbulence forcing it towards the flame, would not normally ignite so readily, but under higher temperature and pressure conditions the fuel and oxygen molecules are brought closer together and it burns quite cheerfully.

In theory the result is a faster, more complete burn, possibly with a mixture which, overall, is leaner than the conventional optimum. Experimentally the results show up as an ability to retard the ignition without losing torque; possibly an increase in torque or the use of a weaker mixture for the same level of torque.

Carburettor theory

A carburettor works by manipulating gas energy; any gas has various energy levels – one due to its motion (kinetic energy), one due to its height or head above some datum level and one related to its pressure and density. Bernoulli's theorem states that, as long as nothing is added to or taken from the gas, then its total energy level will remain constant. This implies that if one of the energy levels is raised, then one or more of the others must be reduced in order to keep the same total. It can be expressed as an equation:

$$\frac{v^2}{2} + \frac{p}{D} + hg = K$$

where v = gas velocity
       p = gas pressure
       D = gas density
       h = height above datum
       g = gravitational constant
       K = constant

This applies to a perfect gas, which is incompressible and in steady, adiabatic flow. Real air is less than perfect, it is not even one gas, and the equation for compressible gas is a little more complex:

$$\frac{v^2}{2} + \int \frac{dp}{D} + hg = K$$

This allows for a more complicated relationship between gas pressure and density, but the broad effect is the same. If the air does not change height very much, the term hg will be constant and will not affect things. The air outside the carburettor, and inside the float chamber, will be static and will have a certain pressure and density. When it is drawn into the engine it gains kinetic energy – in proportion to the square of its velocity – and the carburettor is designed to raise the velocity locally in the region of the fuel jets. There is often a narrowed section called a venturi to do this. The term p/D must therefore be reduced in proportion and, if we assume that D does not change, this leaves quite a significant drop in pressure, p. This pressure drop, between the float bowl and the carb's venturi, is applied directly across the main jet and main spray tube and the relatively high pressure inside the float bowl pushes fuel up through the jet. If D does change, due to atmospheric conditions, this may well affect the rate of fuel flow.

Fig. 38. An air-bleeding carburettor and the effect on WOT mixture strength caused by increasing the main jet size and then increasing the air jet size

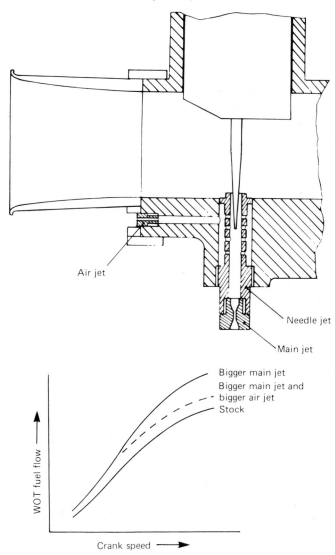

The engine speed and displacement, and the size of the venturi control the air flow and its velocity; the velocity-induced pressure drop and the size of the fuel jet control the fuel flow. Changing the jet will eventually give the required air/fuel ratio.

As the fuel emerges from the spray tube, the second function of the carburettor is to break up the liquid into the smallest possible particles which are light enough to be carried along in the air flow and which can be dispersed fairly evenly throughout the air. This is a function of the air speed and the design of the spray tube.

Having got the mixture right at one speed, if the speed is then doubled the air flow will be doubled (assuming the same volumetric efficiency) and its velocity through the venturi will be doubled. But as the pressure drop is related to the square of the velocity, this would actually be quadrupled and instead of delivering twice the original fuel flow, there would be a strong chance of four times the flow emerging from the spray tube. In practice the fuel flow does not go up quite as dramatically as this (because the density factor changes, there are losses due to shear and friction in the gas and because the gas speed is not constant, it is a series of pulses which are further spaced at low speed, which means that the actual peak speed is further away from our theoretical, constant 'mean' speed).

So the fuel flow does not rise in proportion to the square of the air speed, but unfortunately it still is not a linear function either. If the air/fuel mixture is correct at low speed, it will still tend to be too rich at high speed. The change of fuel flow with engine speed is sometimes called a fuel slope and the object of tuning the carburettor is to make this slope correspond to the engine requirements.

A larger main jet will simply lift the whole slope and make it rise at a steeper angle, that is, get richer at higher speeds. One way to alter the angle is to bleed air into the fuel flow through the jet. This is controlled by an air jet, as shown in Fig. 38, and the flow through all jets is, as described above, approximately proportional to the square of the air velocity. Applying this to the air jet, there will be a small flow when the gas speed is low and the air bleed will reduce the fuel flow, but not by a great amount. At higher speeds the air bleed will flow much more and will cause a much greater reduction in fuel flow; it will make the slope less steep. There are now two adjustable items to play with, the air and fuel jets, and they can be combined to tailor the fuel flow requirements quite accurately.

Table 7.2 shows some figures taken from a real engine in which the fuel slope is virtually linear – proving that it can be done. Unfortunately, the SFC and air/fuel figures show that this still is not quite what the engine wants; the actual fuel slope and the required fuel slope are shown in Fig. 39. This corrected slope is based simply on reducing the fuel flow in order to bring the SFC to a value of around 0.53, assuming the power stays where it is. What usually happens is that as the mixture gets close to the optimum

Table 7.2 WOT fuel flow and air flow

crank speed (N) (rev/min)	volumetric efficiency (E) (%)	fuel flow (pt/h)	SFC (pt/hp-h)	air flow factor (N × E)	air/fuel ratio
3900	90	9.0	0.57	3.51	11.1:1
4900	90	12.2	0.62	4.41	10.3:1
5830	97	14.5	0.57	5.66	11.1:1
6750	103	16.5	0.53	6.95	12.0:1
7700	100	20.0	0.58	7.70	11.1:1
8500	90	22.7	0.66	7.65	9.4:1

The air flow factor is simply the product of the engine speed and volumetric efficiency; multiplied by the engine displacement it would give bulk air flow which, divided by the carburettor cross-sectional area, would give a mean value for the local air velocity.

setting, the power increases slightly which would, in this case, being the specifics down to below 0.53 and may make the engine run too weak for its new power level. Then the process would have to be repeated, but each time it would get closer to the optimum settings.

The air bleed, introduced into the fuel flow by a drilled emulsion tube, also has several useful by-products. It emulsifies the fuel into a frothy mixture which is much easier to atomize and one which will not drop back down the jet too quickly when the throttle is closed. Then when the throttle is opened again, there is less delay before the fuel reaches the top of the spray tube. Also, when the engine is running at a steady, low speed, or

Fig. 39. Comparison of the actual fuel flow at various crank speeds, with the flow needed to give the correct mixture strength (see Table 7.2). The mixture gets progressively richer either side of 6,750 rev/min

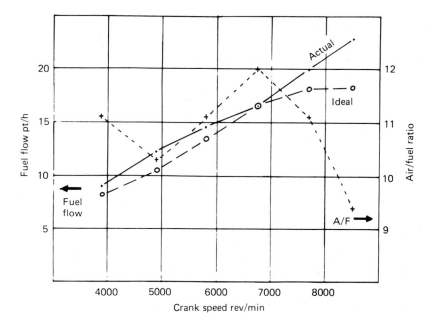

idling, the chamber around the emulsion tube will fill with fuel. If the throttle is suddenly opened, this readily-available, albeit temporary, supply of fuel, is enough to help the engine get through the transient region without any spluttering or hesitation caused by the fuel system not responding quickly enough.

So far, we have only considered a wide open venturi and one simple jet system.

The carburettor also needs some means to control the air flow, independently of the engine's demands – namely a means of throttling the flow. At this point, carburettors divide themselves into two families, slide carburettors and CV – constant velocity – carburettors.

Slide carburettors: these raise and lower an air slide (either flat or round) to throttle the air flow directly above the main spraying system. Lowering the slide, while reducing the bulk air flow, also reduces the cross-sectional area of the venturi and tends to maintain the same sort of velocity across the top of the spray tube. Therefore, to reduce the fuel flow in proportion to the air flow, a tapered needle is attached to the air slide and as the slide is lowered, it progressively restricts the needle jet – which is incorporated into the spray tube. When the throttle is wide open (WOT) the needle is clear of the jet and the fuel flow is dominated by the main jet.

This provides several more adjustable factors; the needle jet size and the rate of taper of the needle (provided that the area of the jet minus the needle is smaller than the area of the main jet); the cut-away on the leading edge of the air slide and the position of the needle relative to the air slide (it is usually provided with five grooves to accept the mounting clip). Within these adjustments there is enough scope to provide adjustment from about $1/8$ throttle to WOT.

The idle system is a separate jet, fed by an air bleed and usually regulated by a tapered screw which restricts either the air passage or the fuel passage. Its main outlet is downstream of the air slide, although it will have one or more by-pass outlets under the trailing edge of the air slide. The by-passes are opened up as the throttle is opened, providing an increasing fuel supply before the main system takes over.

CV carburettors: these instruments have the same type of adjustment and part-throttle settings but differ in having a separate throttle valve downstream of the air-slide/spray tube area. This governs the bulk air flow while the air slide is fitted in a sealed chamber which is vented to the pressure in the venturi. A spring holds it in the closed position.

When the throttle is opened the air flow increases and raises the air velocity in the venturi. The drop in pressure is transmitted to the piston chamber where it causes the piston to rise, until the speed (and pressure) has dropped to the designed level. If the throttle is then opened further, the same

Fig. 40. A constant-velocity carburettor, using a primary and secondary main jet system. The pilot system has its outlet controlled by a tapered needle but it also has a gallery blocked by a plug, which gives access to the (several) by-pass outlets. By removing the plug, any of the by-pass outlets can be blocked and another drilled in a new position, to cure flat spots just above the idle position

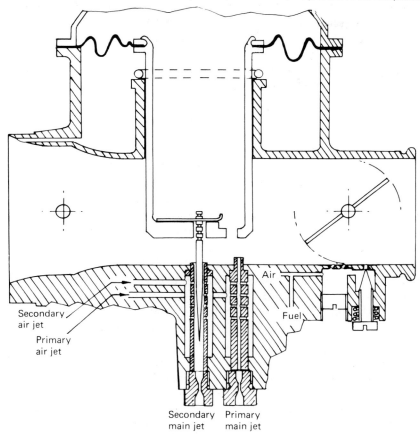

process will be repeated until the air slide is fully open. The effect is to keep a constant pressure drop across the main jet while the needle is controlling the fuel and only to allow the air slide to open as quickly as the engine can manage, regardless of how the throttle is operated. This tends to give better pick-up.

The idle system is similar to that used on slide carburettors, with the outlet and by-passes located under the edge of the throttle valve. As this is usually a butterfly valve, the by-pass positions can be accurately placed to come in at precise throttle positions.

Some CV carbs have two main jet systems, each with its own air bleed. The secondary system is the one described above. The primary system is located downstream, with its spray tube below the trailing edge of the air slide. This provides some progression from the idle system to the main system and can be tuned by altering the fuel jet or the air jet. Like the pilot system, it will continue to deliver fuel at all throttle positions.

It is a lot easier to set up carburettors on a dynamometer using flow meters; the alternative, to use a track, demands great sensitivity from the rider and patience from the tuner and even then it is difficult to get the carburation right in the midrange, because the engine is always accelerating through this region. As it gives a crisper response when the mixture is slightly weak, there will always be a tendency to set it up this way and it is usually worth finding what appears to be the optimum and then going one step further in the rich direction.

For simple power production the slide type has a clear advantage; it can flow more air and (especially the smoothbore type) will respond to resonance in the intake tract. It has a number of disadvantages, particularly for road use or for use by unskilled riders. If the rider arbitrarily opens the throttle, then CV carbs will give an immediate and progressive engine response from any engine speed, because, although the throttle valve is open, the air slide will only rise in proportion to the demand from the engine; therefore the gas speed and the fuel flow will stay within the limits of both the carburettor and the engine.

On slide type carbs, if the rider opens the throttle too quickly at low rpm, the air speed can drop below the point at which it draws fuel through the jets properly; power will fall, the bike will go slower not faster, there may even be a misfire. The rider then has to roll the throttle back in order to accelerate. If the rider opens the throttle too quickly at high rpm, a similar thing may happen, the engine may not get the right mixture of fuel and air and may hesitate, the power suddenly coming in very strongly as the air flow stabilizes. Alternatively the power may come in just as quickly as the rider opens the throttle – quickly enough to spin the back wheel or lift the front wheel off the road.

Slide carbs also have the full air flow pressing against the air slide; consequently each carburettor needs a heavy return spring to prevent the slide from sticking. This either makes the throttle very heavy, or it has to be given a large mechanical advantage, which means that it has to have long travel. CV carbs only have to operate butterfly valves with very light return springs, so quick action twistgrips can be used, going from closed to wide open in about a quarter of a turn.

Carburettor tuning

While each sytem within the carburettor is meant to control a specific region in the load/speed envelope, there is inevitably a lot of overlap. Changes to one system may easily affect its neighbours and so the best procedure to follow is one which selects the major items first and finishes with the least consequential. In most carburettors this will be:

1. Float height
2. Main jet and main air jet
3. Idle system
4. Needle jet/needle taper; repeat main jet
5. Air slide
6. Idle system
7. Needle groove
8. Repeat process

Float height: the float height is not a dominant factor but it must be the same on all carburettors and once checked it should remain unaltered. There is no reason to deviate from the stock settings, although if an engine has been drastically modified and the carburettors are not responding properly to adjustments, it is worth checking that the fuel tap, line, line filter and needle valves can cope with the required flow. As an approximate check, the engine will require a flow of 0.6 × the peak power, measured in pt/h, plus a small safety margin, and the actual flow can be measured by running the fuel through into a measuring cylinder and timing it. If a much larger main jet is being used and there are signs of fuel starvation during acceleration, it may be worth increasing the fuel level (see needle valve below).

During braking and acceleration the fuel will swill to the front or back of the float bowl, respectively. This should not affect fuel supply to the main jet, which will be placed centrally, but it may cause fuel to flood up through the pilot jet, causing a temporarily rich mixture which can spoil subsequent pick-up and acceleration. This is often a problem on carburettors fitted at a steep downdraught angle when fuel can flow forward/up through the main/needle jet.

Main system: this often feeds all of the other systems, but its first function is to give the correct WOT mixture all through the speed range. The safest way to do this is to start with a main jet which is too rich, and work down until the mixture is correct at the bottom of the speed range, and then alter the air jet to get the correct mixture at peak speed. If there is a choice, use the largest possible air jet, as this will give better atomization. On most production machines the air jet is either a brass bush pressed into a passageway or is a simple drilling in the body of the carb. In order to make it adjustable it is necessary to remove the original bush, tap the hole with a 5 or 6 mm thread to take suitable jets from a different type of carburettor; alternatively conversion kits are available from performance shops. The range of air jets normally used is 0.4 to 2.0 mm.

There are several different types of main jet (as shown in Fig. 41) and they have different flow characteristics and are even sized differently. Some are calibrated in cc/min and some are simply measured as a diameter of the jet,

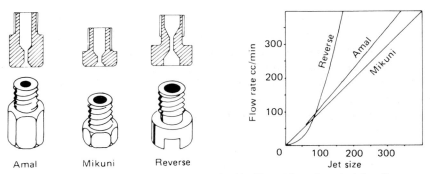

Fig. 41. There are several types of main jet, each with different flow characteristics. Some are calibrated by flow (in cc/min), others by jet diameter (in mm)

in mm. Even jets of the same type from different manufacturers may not exactly correspond in their flow rates, so it is important to identify the type and stick to the same make when adjusting the carburation.

Occasionally when a much larger than stock main jet is being used, the flow rate no longer responds to jet changes. This is because the final portion of the needle in the end of the spray tube is causing a greater restriction than the main jet. To check, measure how far the needle extends into the spray tube on WOT, then measure that part of the needle with a micrometer and the corresponding part of the spray tube with a small gauge or a suitable drill bit. The area available for fuel flow is $\pi(D^2 - d^2)/4$, where D and d are the two diameters. This area needs to be significantly more than the area of the main jet, $(\pi d_j^2)/4$, where d_j is the diameter of the jet. Alternatively, see how the engine runs when the main jets are removed. If the spray tube area is not bigger than the main jet then it will be necessary either to use a smaller main and a smaller air jet, or to find a new combination of needle and needle jet.

If, on the other hand, the engine is unduly sensitive to jet changes – going from fully rich to fully weak within a few jet sizes – this is usually a sign that something else is wrong such as a mismatch between intake and exhaust dimensions, wrong carburettor size, wrong type of spray tube, etc.
Idle system: the individual adjusters should be set to give the best idle (i.e. the highest idle speed for the lowest throttle stop setting), which can be quite a delicate operation on a four-cylinder machine. The alternative is to start at the stock setting (usually something like 1½ turns out) or to use a Colortune plug. The carbs should also be balanced to get the smoothest possible low-speed running. In its crudest form this means checking that all throttle valves open together and can be set by trapping pieces of welding rod under the valves and adjusting the throttle stops until they are all released together. Once the engine is running it is easier to use vacuum gauges to get the same readings on all intake stubs.

Most four-cylinder engines have the carbs coupled in pairs and one carburettor is used as a reference – its adjuster is not touched (some models do not have an adjuster on this carburettor). Suppose the reference carb is No. 3. The balance method is to use the throttle stop to equalise 4 and 3; then synchronize carbs 1 and 2; finally use the central adjuster located between carbs 2 and 3 to equalize both pairs 1/2 and 3/4. The main throttle stop is then used to regulate the idle speed.

One problem on tuned engines is that they often develop a flat spot just off idle and sometimes the only way to cure it is to enrich the idle systems, at the expense of an even tickover. Some machines have the pilot jet blanked off inside the float bowl and it is fed via a drilling from the main jet. This starts up flow through the main jet, making the transition easier when the throttle is opened. It also prevents the pilot system being swamped and possibly gassing up the engine when the machine is braked hard. As the idle system flows fuel at all throttle settings, it can have some effect on high-speed carburation.

Primary main (where fitted): this is a jet with its own air bleed and an outlet positioned under the trailing edge of the air slide or piston. As the throttle is opened, fuel is fed from the pilot and its by-passes and, as the air speed increases the piston is still held down and so a very high local speed is produced over the primary main. This then begins to supply fuel, filling in the gap between the pilot and the secondary main. As the piston lifts, the air velocity over the jet is reduced and its contribution becomes less significant, although it continues to supply fuel. Its major effect is in the $\frac{1}{8}$ to $\frac{1}{4}$ throttle region and it can be tuned in the same way as the secondary main, by changing jet and air jet sizes.

Its effect can also be altered by changing the way in which the piston behaves. A stronger spring, for example will hold the piston down longer, increasing the air speed over the primary and making it supply much more fuel in the low throttle positions without affecting its performance once the piston has lifted.

Conversely, if a spacer is fitted under the piston, to raise it slightly, the gas velocity over the primary main will be reduced, weakening the mixture at low throttle openings. This may be useful if a larger primary is used in order to produce extra richness at wider throttle openings, but is not required to change the mixture at low throttle openings. Although it only contributes a small portion of the total fuel flow, the primary main continues to flow at all throttle positions.

Air slide: the cutaway on the front edge of the airslide affects the mixture at low openings, up to $\frac{1}{4}$ throttle, and a smaller cutaway gives a richer mixture. The number code is usually the height of the cutaway, in sixteenths of an inch, or on Dell'Orto carburettors, in tenths of a millimetre.

Piston valve: in CV carburettors the air slide is replaced by a sealed piston with its top surface vented to the low pressure in the venturi. This lifts the piston against its weight and the force of a light spring. Holding the piston down, by using a stronger spring, will increase the air velocity under it and reduce the pressure over the fuel jets. It will not necessarily cause more fuel to flow from the main jet because it also holds the needle down. Raising the piston by fitting a spacer under it or by using a weaker spring will reduce the air velocity but will also raise the needle. If CV carbs are mounted at a steep downdraught angle then the inertia of the air slides during braking can lift them, upsetting the carburation if the throttle is opened immediately after braking is finished. There is no cure for this apart from changing the type of carb or its position. The effect it has can be reduced by using a stop-light switch in the brake to switch off the fuel pump, so no fuel is supplied to the carbs during braking.

In other circumstances the air slide can flutter, the movement producing unpredictable effects on the carburation. This can usually be prevented by changing the size of the hole which vents the underside of the slide. This can also be used to tailor the carb's response to throttle movement – a larger hole will make the slide lift earlier/faster, a smaller hole will reduce the rate of movement, acting as a damping mechanism.

Needle jet: in most motorcycle carburettors the needle jet holds the main jet and forms the emulsion tube and spray tube. Its diameter at the upper portion also regulates fuel flow in the ¼ to ¾ throttle positions, as the tapered needle is withdrawn from the jet. The effective jet size is a function of its diameter and of the size of the needle, so the two must be selected as a pair. Throughout most of the part-throttle range the needle jet gives coarse adjustment, while the needle gives fine adjustment. Mikuni needle jets are size coded with a letter and a number. The letters progress in 0.05 mm steps, so N is 2.55 mm and O is 2.60 mm. The numbers represent steps of 0.005 mm, so N-0 is 2.550 and N-1 is 2.555, etc., although usually only even numbers and 5s are available.

The height and the shape of the spray tube also have an effect on the atomization of the fuel and its delivery to the engine.

Dell'Orto needle jets follow the same pattern; they have primary choke type atomizers (usually used on two-strokes, but fitted to some four-stroke applications) and several types of emulsion tube. The height of the spray tube affects mixture strength, the taller version giving a weaker mixture at low throttle openings. The emulsion tube which has holes at the bottom runs richer at part throttle because the fuel fills the well surrounding it. Those with holes at the top allow the air jet to affect part throttle operation, and run weaker.

The jets have a code number, e.g. 9979 × 28, which refers to the carb body (PHBH in this case) and a jet type, e.g. BC, which is stamped on the jet and identifies the overall height, the height of the spray tube and the emulsion tube pattern. Finally the size of the needle jet is given in hundredths of a millimetre, e.g. 262 represents a diameter of 2.62 mm.

Amal jets are available in emulsion tube or primary choke form and have a simple bore size to identify them.

Needle: this has to be selected in conjunction with the needle jet and the diameter at any given point determines the fuel flow for that amount of throttle (or piston lift, in the case of CV carbs). Fine adjustment is provided by having five grooves machined in the top of the needle to accept the locating clip. The grooves are numbered from the top, number 1 giving the weakest setting.

Mikuni code their tapers in quite a complicated way, because not only do they have different lengths of needle to suit different carburettor sizes, but some needles have two tapers. As an example: 5GN36-3.

5 – the series number (3, 4, 5 or 6 – also determines the length)
G – indicates the angle of the upper taper
N – indicates the angle of the lower taper
36 – factory reference number
3 – the last digit is not the part number but refers to the groove in which the needle is located for specific models.

The letter code uses every letter of the alphabet, starting with A which represent 0 deg 15 minutes; each letter goes up in 15 minute steps, so G represents 1°45' and N represents 3°30'.

Note that a larger needle jet will enrich the mixture everywhere up to ¾ throttle or a fraction further, while a change of taper can, if necessary, richen the mixture below half throttle and weaken it above this position, or vice versa.

Some needle mounts have a plastic spacer which tips the needle to one side, making it bear against the side of the needle jet. This is to prevent the needle fluttering in the pulsating air flow.

Dell'Orto needles have four notches, No. 1 being at the top, and they are sized according to the diameter of the parallel section, the diameter at the tip and the length of the tapered portion. A few have a double taper, identified by their height and the diameter at the change of taper. It is possible to find tapers which are richer at the top end and weaker at the bottom end, or vice versa, allowing a full tuning process; the stockist will be able to identify the sizes from his code chart.

The needles have a single letter and number, e.g. K6, the letter identifying the carburettor body, while the number relates to the dimensions of the needle itself.

Amal carburettors usually have only two or three optional needles.

Pilot jet: this controls the idle and the fuelling up to about ¼ throttle, although it continues to flow throughout the entire throttle range.

These are the major systems, but there are several other features, some of which are optional but can be fitted to stock carburettors.

Needle valve: this regulates the flow to the float chamber and, if an engine is drastically modified, it may not be able to flow enough fuel. Larger sizes are available.

The needle and its seat must be in good condition – and even when they are, the valve is susceptible to flooding particularly if there is dirt in the fuel or if there is a fuel head of much more than a few inches. A line filter should be used and, if a pump has to be used, a pressure regulator or a header tank should be incorporated. The needle valve should not be greater than 1.3 × the main jet size (gravity feed). It should be smaller than the main jet if a pump is used, with a return to the tank containing a restrictor jet which is smaller than the needle valve.

Fuel line: to cope with increased fuel flow, large-bore taps are available. Many machines use vacuum-operated taps which depend on a tapping just downstream of the carburettor. There is not much vacuum on WOT and if the air filter is removed there will be even less – possibly not enough to hold the fuel tap open. This can cause fuel starvation, but only after the motor has been held wide open for more than a few seconds. It can be tested by measuring the vacuum needed to open the fuel tap (and measure the fuel flow from the tap) and comparing this with the vacuum available at the intake stub on WOT. If a water manometer is used for this, it will need a shut-off valve or some other device to prevent the liquid being drawn into the engine when it is closed down to part throttle. The PRIME position on some fuel taps does not provide enough fuel flow to support WOT running. Fuel starvation can also be caused by blocked or inadequate fuel tank air vents.

Main jet baffle: a circular fence which fits around the main jet and prevents fuel being swilled away from the jet during braking, accelerating or when the bike is used over bumpy ground.

Main jet extension: a spacer which fits between the main jet and the needle jet so that a longer needle can be used. May also include a baffle.

High speed jet: an extra fuel system, with its own jet and spray tube, fitted into the carb body upstream of the throttle valve. The position of the spray tube means that air will not flow past it until the throttle is opened beyond a certain position – usually ½ to ¾ throttle, so the jet does nothing below this point. On WOT it supplies an increasing amount of fuel as the engine speed increases, thus raising the top end of the fuel slope. It can be regulated by changing the jet. Also called 'power jet'.

Accelerator pump: a similar arrangement to the high-speed jet, with its own spray tube, this system also contains a small, sprung plunger which is moved by a linkage attached to the throttle linkage or to the airslide on some Dell'Orto carburettors. As the throttle is opened, it pumps one stroke of fuel through the spray tube, providing an immediate richness if the throttle is snapped open. On some layouts the system continues to work as a high-speed jet (see above).

Vacuum valve: there are several applications where diaphragm-type valves are opened or closed by engine vacuum, against a spring, usually to bleed air to the intake when the throttle is closed and the engine is on the overrun.

Vacuum bleed: low pressure in the venturi is sometimes vented to the float chamber, to reduce the pressure there (and weaken the mixture) or to operate a valve which regulates the pressure in the float chamber (as an altitude control, for example).

Second air jet: on some carburettors there is an extra air jet, whose entrance is blocked by the throttle valve, being uncovered when the throttle is wide open. In this way two air jets can be used at WOT but only one is used for part-throttle positions.

The rest of the intake system also has an effect on the carburation, as it can cause a pressure drop which will affect the pressure at the fuel jets and some parts can alter the engine pulses which travel along the intake tract. The position of the spray tubes in relation to the passage of these pulses can have quite weird effects on the fuel delivery.

Air box: this acts as a surge tank which can damp out unwanted pulsations or, it seems, can strengthen beneficial pulsations. Often the air flow and power are increased by the presence of the air box, particularly in the midrange, and sometimes at peak power. To achieve this, the air box needs a very large volume; say 7 litres minimum for an engine giving 100 bhp. The ZXR750 Kawasaki has a 12-litre air box on an engine making 103 bhp, although the race tuned version is capable of about 140 bhp. This is pressure fed on the L and M versions, as are several other machines, like the ZZ-R1100 and the fuel injected Ducati 851/888. Note that the air intake in the front of the fairing does not need to be huge (the engine uses one carburettor at a time). On a machine capable of 175 mph, a forward facing intake will pressurize the air box if it is bigger than 14.8 cm^2 (2.3 in^2). The air pressure, even at this speed, is not enough to make big differences to horsepower; the maximum available is about 16 inches of water (about 0.6 psi (0.04 bar) which, in terms of boost, is not much but it is better to receive it than to reject it; it is worth something like 3% more horsepower). While 0.6 psi won't do much in the way of supercharging, it can do a lot to upset carburation. To compensate for these changes, the float bowls should be vented into the air box. The air pressure will change at various positions, depending on its speed – at the narrowest part of the intake duct the air will be travelling fastest and will have the lowest pressure, the pressure will be highest where the air speed is lowest.

The positioning of the air box and the length of the carb intakes extending into it are critical and must be found by experiment.

On a stock machine the airbox has several other functions; it holds the air filter (see below), the outlet to the crankcase breather (which must be re-routed into a catch tank or into a new filter housing if the air box is removed), and the intake silencer.

Sometimes the entry to the air box or the silencer itself is restrictive at very high rates of air flow. Removing it, or making the entry larger, will possibly allow the engine to flow more air. It will almost certainly reduce the pressure drop – and effectively raise the pressure at the carb's venturi, making the mixture run weak. This effect will be much greater at high speed than at low speed, and greater at wide throttle than at small throttle openings, so the result will be that the fuel slope is made too shallow. With engine outputs exceeding 85 bhp and speeds over 140 mph, breathing can be a problem even with a still-air box fitted. The entry to the air-box, or to the bellmouths, may be in a chamber formed by the underside of the tank, fairing, seat, etc., and the air flow over the sides of the bike may put this chamber in a low-pressure region. Alternatively, the entry to the chamber may become restrictive. Several stock machines, as well as racers, now duct cold air to this region from intakes in the front of the fairing.

Dynamometer tests may not reveal this problem, unless the tank or fairing is causing an obstruction, and anyway tests are often run with the tank removed. A restriction at the airbox, chamber, or entry to the carbs, will increase the depression at the carburettors making the mixture too rich. A low pressure which is caused aerodynamically will reduce the air flow and power, but any mixture changes it causes depend on where the float bowls are vented. If they vent to the same low-pressure region, then there will be no change in mixture strength, just a reduction in air flow. A test for this condition could include fitting long hoses to the vents and moving them to a still air region, or even a slightly high-pressure region.

Air filter: While a well-designed air filter will not be restrictive to air flow – at least up to about 90 bhp – it will still be more restrictive than individual filters, because they have a greater surface area in total. Even if the change is not enough to make much difference to the bulk air flow, there may well be enough of a pressure difference for the carburettor to detect. Once again the fuel slope will be altered. The use of individual filters usually means that the air box cannot be fitted and they may also alter the length/shape of the entry to the carburettors.

Trumpets (intake stacks): a smooth, widely-radiussed bellmouth improves air flow into the carburettor, either by straightening out the air flow or by making the carb less sensitive to engine pulses. Where there is a problem with fuel 'stand-off' or spray back from the carburettor, then changing the length of the trumpet or stack will usually cure it.

Throttle and choke plates: where these valves are mounted on spindles, they obviously give some restriction to the gas flow, even when completely edge on. If a cold-start valve is fitted (most carburretors have a cold-start jet instead) it can be removed (and a temporary cover placed over the carb entries for cold starting) and the throttle plate ground to a knife-edged streamline shape. Make sure that the valve is completely edge on in the WOT position.

Carburettor type: for the same section area a slide carburettor flows more than a CV carb – by some 20 per cent or more – and a smooth-bore carb flows about 5 per cent more than an equivalent-sized conventional type. A flat-slide carb has better air flow under its edge at part-throttle, allowing better control of the part throttle carburation and better pick-up. The smooth-bore types, with the minimum of steps and surface irregularities are also more receptive to pulse tuning in the intake, and will show larger increases in air flow when the intake length is resonant with the engine frequency.

Surge tank/cross porting: fitting a reservoir downstream of the carburettor can absorb pulses when the intake valve closes; alternatively they can be

Fig. 42. A large, properly-designed airbox can actually increase air flow. Most, like this unit on the Yamaha FZ750, have individual bellmouths extending into the box

Fig. 43. The air box can also be a source of cool, slightly-pressurized air, as on this Kawasaki ZX1000-Al. The two intakes are insulated from engine heat and lead to the high-pressure region at the front of the fairing. Honda used a similar system on their V4 racers while the later ZZR and ZXR Kawasakis had much larger, fully pressurized air boxes with the float bowls vented into the air box entry duct.

Fig. 44. The Ducati 851's Weber-Marelli fuel injection has an air-flow sensor which automatically compensates for changes in flow owing to the forward motion of the bike. These two intakes have as much area as an engine needs to make well over 100 bhp – with minimal disturbance to the aerodynamics.

diverted to a cylinder whose valve is open. Experiments suggest that this has more effect at low speed or small throttle openings, possibly because the interconnecting pipe is not large enough to make a significant change in flow at high speed. On a 360-degree Kawasaki twin, a 0.5 in cross pipe (equivalent to about 10 per cent of each intake area) halved the WOT fuelling below 6,000 rev/min, with no appreciable change in load. During quarter-throttle road loads, it reduced the fuel flow from 43 to 38 cc/min and in quarter-throttle acceleration tests it halved the time taken to go from 2,000 rev/min to 5,000 rev/min.

Fig. 45. The choice of replacement filters. Both the gauze and the foam types filter more effectively when they are oiled

Fuel injection

The development of microprocessors has allowed good enough control to make fuel injection a workable proposition. Typically the system works by running a fuel gallery at a constant pressure, from which injector nozzles squirt fuel into the intake tract, close to the valve. Fuel delivery is controlled by regulating the time which the injector is open. The air flow is throttled by individual valves.

Fuel is pumped from the tank to the gallery and a pressure regulator senses the pressure drop between the gallery and the intake tract (allowing the system to work with supercharged engines), opening when necessary to return excess fuel to the tank.

The processor makes a primary calculation, depending on engine speed and an air flow meter or the throttle position, and this determines the main fuel quantity/injection time. A secondary calculation, based on data from transducers which monitor the engine temperature, air temperature,

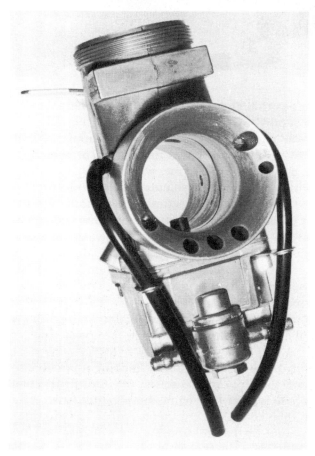

Fig. 46. Amal's smooth-bore carb, showing the 'jet block' which maintains the contours of the venturi, leaving a narrow slot for the air slide to pass through

Fig. 47. Mikuni smooth-bore carburettor

Fig. 48. Mikuni's flat-slide carburettor is also smooth-bored, except for the narrow underside of the air slide. The nozzle just to the right of the air jet is connected to an accelerator pump

starter switch, throttle idle position and, in supercharged engines, sensors which detect knock and boost pressure, is used to fine tune the fuel delivery. There may also be an ignition control which regulates the timing.

The controls conform to a pre-determined fuelling map, which is programmed during engine development. The message is translated into a pulse of electrical power which energises the solenoids in the injectors for the required amount of time. On all systems the injectors operate once per intake stroke; on some they operate once per revolution once the engine has reached a particular speed.

There is no scope for adjustment built into these systems, although this does not mean that this is not possible. It can be done by either re-programming the processor, by altering the pressure in the fuel gallery or by changing one of the correction factors.

Changing the program is the most satisfactory way as the other methods give fairly coarse adjustments, which will apply through the entire range of speeds and loads.

Specialist firms will modify the pressure regulator, making it possible to adjust its spring pre-load and therefore to adjust the pressure in the fuel gallery. The other easy way is to replace the engine temperature sensor by a variable resistor, which can then be used to enrich or weaken the mixture.

A better solution is to exchange the control unit for one which can be programmed, for example an EPROM chip containing data for the complete fuelling and probably ignition maps – which can be edited by connecting the unit to a computer keyboard and altered while the engine is running on a test bed. Thus the settings for any speed and load can be swung through the full range and the optimum position found simply by pressing a few keys and watching the reaction on the dyno's instruments.

The reason that fuel injection is not widely used in racing is that it doesn't give more power than carburettors and it does entail a fair amount of electronics, fuel pumps etc., which must be protected, screened and supplied with electrical power. The engine air intake, like a carburettor, must contain either a butterfly valve or a slide valve, with the same options (smooth bore v. restriction, light v. heavy return springs). Ultimately there is no gain in air flow and no increase in power.

Where fuel injection may give better power is (a) in installations where there is not much room (the injector is much more compact than a carb) for instance in V-motors, (b) where there is a heat soak problem which causes fuel vapour locks in float chambers (again in V-motors), (c) in transient conditions where it is easier to adjust the injection system and therefore requires less skill/time to get better pick-up and throttle response and (d) by using the same sensors to regulate other components such as ignition timing, traction control, variable exhaust geometry, anti-knock measures and so on, which would mean that the engine could be allowed to run closer to the edge of reliability or of controllable power, where knock sensors, rev limiters and devices to sense wheelspin could prevent the rider from hurting himself or the engine parts.

The success of Ducati's 851/888 variants with Weber Marelli injectors shows that there is plenty of potential and, as there is little experience of using fuel injection on motorcycle installations, there should be plenty of room for development in injector nozzle patterns, the positioning and timing of the injector, types and dimensions of throttle valves and so on.

Nitrous oxide

As this is not a fuel it can be used legally in road vehicles. It is an oxidising agent, stored as a liquid under very high pressure and supplied via a solenoid valve to a spray tube downstream of the carburettor. When the valve is opened, the nitrous oxide is delivered to the intake and immediately turns into a gas, its latent heat of vaporization having an enormous cooling effect on the intake gas and the engine. Once inside the cylinder, under compression, it gives up free oxygen which can be used to burn more fuel – which has to be supplied separately.

Some systems are quite crude, simply delivering more or less random quantities of fuel and nitrous oxide. Others have a valve which operates the nitrous oxide and the fuel supply together, and which uses the high-pressure

Fig. 49. These Keihin semi-flat slide instruments are meant to combine the best of the flat-side characteristics with the properties of CV carburettors

gas supply to atomize the fuel spray. A jet in the fuel line regulates the fuel to suit the amount of gas being delivered.

The system can be staged, using two controls to bring in the full amount of nitrous oxide, but there is, so far, no finer degree of control. This is unfortunate for many applications because it is not difficult to boost power by 50 per cent – which makes for quite a violent transition.

The ultimate restriction is caused by knock, although the engine can tolerate unusually high gas pressures before knock begins, possibly because of the cooling effect of the gas and possibly because it tends to cushion combustion. The other limit is the useful operating time; a gas bottle which is a convenient size for installation on a motorcycle will last for about 40 sec at an output of 100 to 130 bhp, although the duration will be longer at lower power levels.

Chapter 8

Ignition

Despite the many types of ignition system, there are two main features which are important to the engine; the quality of the spark and its timing. The spark must have enough energy to start the gas burning; now a certain amount of power is fed to the ignition system during the time that it is not delivering the spark (the rise-time of the system, or the closed dwell) and this available energy is released in one short burst. The input is limited – by the system voltage, the resistance of the coil's windings and the time available – which diminishes as the engine speed increases. The output – the spark – has voltage, current and time, or duration and cannot exceed the input for each cycle. It is necessary to reach a certain voltage across the electrodes of the plug before the spark will occur at all. If it is necessary to raise the voltage further, then the current or the duration will have to be reduced.

Different types of ignition system strike a different balance between these three factors. Contact breaker systems have a relatively slow switching speed, which limits the output voltage. The closed dwell is also limited, compared to electronic systems, and this restricts the input at high speeds. There are high-performance coils which will deliver the output needed by a high-speed engine but they tend to take a fairly high current.

Capacitor discharge provides the greatest output voltage and is good on engines which are likely to foul spark plugs, but the spark tends to be of short duration and can lead to misfiring under difficult combustion conditions. On multi-cylinder engines with numerous and often lengthy HT leads, there is an increased risk of external arcing or tracking, or induced sparking between neighbouring leads.

Most stock four-strokes have inductive discharge systems, which produce the most favourable characteristics in terms of spark power and duration. The components are often engineered down to the bare minimum required by the standard engine. When the output requirement is raised by tuning the engine, the system cannot cope.

The ways in which the voltage requirements vary are listed below and the symptoms of a sub-standard ignition system are:

1. A misfire which starts at peak torque.
2. A loss of high-speed performance, or ragged and inconsistent power production at high speed (the bhp graph follows a jagged path).
3. The engine fails to respond to further development.

The system can be tested by lowering the voltage requirement – by using a different type of plug, for example – or by raising the input. One easy way to do this is to increase the system voltage by 2 V and repeat the engine tests. If these steps remove the symptoms, then the ignition system has reached its ceiling.

Sometimes a change to high-performance coils, with suitable solid-cored HT leads and competition plug caps, will be enough. If not, it will be necessary to use an uprated amplifier circuit and this may need to be matched to suitable trigger units. There is a variety of triggers available – including Hall effect triggers, optical triggers, magnetic triggers and pulser coils. The best characteristics for any particular installation can only be found by experiment – or by observing what works on other, similar, engines. Roadsters must have radio noise suppressors fitted to the HT system.

Some of the ignition specialists who supply competition systems can advise on particular set-ups and will be able to build the necessary characteristics into their circuits.

The second feature of any ignition system is its timing. At some stage during the development of an engine, the optimum ignition timing will have to be found – and this may vary with different speeds. If an engine is held at a steady speed on a dynamometer and the ignition timing is progressively advanced then the load will increase steadily up to the point at which the engine knocks. The combustion temperature will also increase. Retarded ignition timing will give a lower temperature, and keep the engine away from the knock region, but it will reduce power slightly and, more noticeably, will seriously reduce throttle response and pick-up. The compromise is to use the minimum advance necessary to get the best load and response. If knock is a problem at a particular speed, then the timing may need to be retarded a shade further, if it is a problem over a wide range of speeds then the combustion chamber or compression ratio will have to be revised (or a higher octane fuel used).

Because the best timing will not be the same at different engine speeds, an advance curve needs to be incorporated in the ignition timing. This can be speed-sensitive (using a mechanical advancer with bob-weights thrown out by centrifugal force, or electronic, where a pulse is generated and the cut-off voltage is reached more rapidly at higher speeds), or it can be load-sensitive. The latter system, mainly used on car engines, usually depends on the depression in the intake stubs. Where a microprocessor controls fuel injection so the ignition system has digital control, a pre-programmed ignition may also relate engine speed and air flow to the timing and can use knock sensors to retard the ignition as needed.

Sometimes the original advancer mechanism can be used; most of the aftermarket ignition system either have fixed timing or an automatic advance curve which is not adjustable. Consequently it is often necessary to find the best timing for peak power and suffer incorrect timing at lower speeds. As the combusion temperature and knock are closely related to the timing, it must be set accurately. Many big four-strokes run slightly retarded for safe, high speed running and advancing the timing by a few degrees gives them a better response and better midrange power. There are aftermarket ignition pick-ups which are offset by several degrees, or a modified pick-up could be made fairly easily to give slightly advanced timing. This will raise the combustion temperature, though, and should not be used on engines which will be held at high speed for long periods.

Knock

There are two malfunctions which cause light, half-engine speed knocking, pre-ignition and detonation. Both will cause piston, valve or plug failure eventually and the higher the state of tune, the less time it will take.

In normal ignition, the burn is started by the spark plug and spreads evenly through the gas. In pre-ignition, something else starts the burn before the spark does; it is usually something which has overheated like the edge of a valve, the plug electrode or some part of the head or piston. Protruding parts, especially those with sharp edges, have a greater surface area (to absorb heat) and less metal under them (to conduct heat away) and as the heat path is relatively poor they run at a higher temperature than surrounding parts. Such a sharp-edged protrusion is likely to cause pre-ignition. The knocking sound is either caused by the expanding gas hitting the piston while it is still travelling upwards or by the pre-ignition flame front colliding with the properly-ignited flame front. The pressure can rise locally to such a level that detonation is caused in the remaining unburnt gas. Pre-ignition causes a rise in combustion temperature which can be enough to cause mechanical failure.

At certain combinations of temperature and pressure, the fuel/air mixture will ignite on its own and, instead of an orderly flame passing through the gas, each particle burns simultaneously. The resulting explosion is particularly violent and is capable of eroding or melting valve and piston material. The fuel's limiting conditions for knock are expressed in its octane rating – the higher the rating, the more knock-resistant the fuel is. Four-star pump petrol varies around the 97 octane mark while Avgas is available up to 115 octane.

Detonation usually happens at low speeds where the trapping efficiency is high, or at peak torque where the air flow is high. It is less likely to occur at maximum engine speed because the volumetric efficiency drops off, the cylinder charge is less dense and there is less time for it to happen (or perhaps it does happen, but without any noticeable consequences).

It can be prevented by using higher octane fuel or by reducing the temperature or pressure conditions inside the cylinder. Usually this means retarding the ignition, making the mixture richer or reducing the compression ratio – whichever causes the smallest power loss. There are alternatives, such as water injection, water/alcohol injection or using fuel additives to raise the octane rating.

Where failures occur, those caused by pre-ignition will probably show signs of melting – deformed piston crown, sometimes like a molten pool with a hole at the bottom of it, severely eroded plug electrodes, etc. Detonation looks like a more rapid failure, often as if a particle of metal had been bouncing up and down on top of the piston, small pieces broken away, components cracked or looking as if they had been bead-blasted.

Spark plugs

According to NGK, there are many factors which govern the voltage requirements of the plug, most of which will be altered to some extent when the engine is modified. Fig. 50 shows how the voltage and temperature vary with the following:

1. Air/fuel ratio. The voltage needs to be higher when the mixture is chemically correct, lower when it is enriched slightly, which also causes the plug temperature to fall.
2. Cylinder pressure. The voltage rises in proportion to the pressure, and so does the plug's temperature.
3. Ignition timing. Advancing the timing will lower the voltage (because the pressure will be lower when the spark occurs); it also raises the temperature.
4. Electrode temperature. The higher the temperature, the lower the voltage (see plug heat range).
5. Plug electrodes. For the same reason that sharp-edged parts run hotter (see above), they also build up a higher electrical charge at the sharp edge, so sharp-edged and thin electrodes require a lower voltage to produce a spark. When plugs are worn, the electrodes become rounded at the edges and are less efficient.
6. Load/speed. For a given load, the plug temperature will increase with speed.
7. Plug type. The ability of the plug to conduct heat away from the electrodes will determine the temperature they run at, as will the contact between the washer and the cylinder head and the tightening torque of the plug.
8. Electrode gap. Increasing the gap raises the voltage requirement.

As well as being made in a variety of sizes, there are different spark plug constructions, which give the plugs individual characteristics. The differences, compared to the standard, conventional plug, are described below.

Fig. 50. Spark plug requirements; variation in spark voltage and plug temperature with other operating conditions

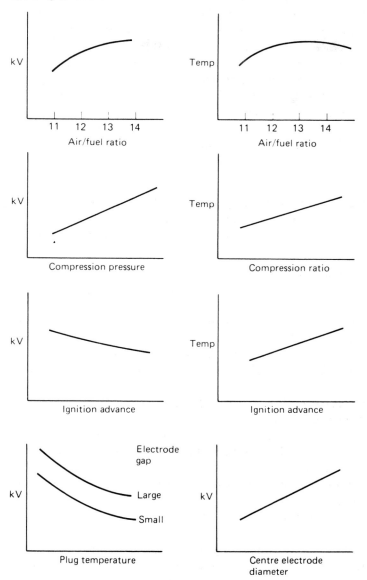

Projected nose: the extended centre electrode and insulator allows the tip to run hotter and to follow engine temperature changes more rapidly. This prevents fouling during cold starts and long idling periods. The plug cannot cope so well with thermal shock or with a progressive build-up of deposits.
Thin electrode: lower voltage requirement gives better performance in borderline conditions. Greater wear rate is countered by the use of

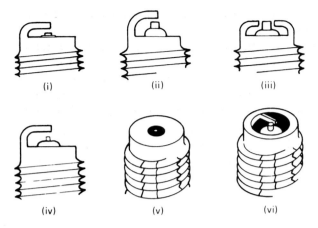

Fig. 51. Types of spark plugs: (1) standard; (ii) extended nose; (iii) double ground electrode; (iv) thin electrode; (v) surface discharge; (vi) racing

expensive materials (e.g. gold palladium) although NGK produce a competition version (suffix EG) which does wear faster, but is cheaper.

Dual ground electrode: able to cope with high temperatures and still have good wear properties.

Surface discharge: runs very cool and needs to be used with a very powerful ignition system which is capable of firing a fouled plug.

Resistor plug: incorporates a high resistance (in the order of 5,000 ohms) to suppress radio frequency noise.

Racing plug: the centre electrode is retracted and the ground electrode is as short as possible, to protect the plug from overheating and mechanical shock. The plugs are usually made from special alloys.

The ability of the plug to conduct heat away from its electrodes it what determines its heat range. This is coded into the plug's type number – a 'cool' running plug will, in the same engine conditions run at a lower temperature than a 'hot' running plug. So, for a particular set of engine conditions, the correct grade of plug must be selected.

If the electrodes run at too low a temperature, oil and fuel deposits will build up on them and the plug will eventually foul. At higher temperatures, the deposits are burned off and the plug goes into a self-cleaning region in which its spark and wear efficiencies reach their highest values. At still

Table 8.1 Spark plug tightening torque – alloy head

plug type	thread dia. (mm)	tightening torque kg-m (lb-ft)
flat seat (with washer)	18	3.5–4.0 (25–32)
	14	2.5–3.0 (18–22)
	12	1.5–2.0 (10–15)
	10	1.0–1.2 (7.2–8.7)
Conical seat (no washer)	18	2.0–3.0 (15–22)
	14	1.5–2.0 (10–15)

higher temperatures, the electrodes wear rapidly and they may get hot enough to cause pre-ignition.

Given the different types of plug and the temperatures created by different engine conditions, it follows that it is not at all easy to 'read' spark plugs. The colour of (and the deposits on) the electrodes and centre insulator do change slightly with operating conditions – particularly with temperature. But unless there is a fault, the changes are very slight and it is necessary to keep the plug from the first test in order to compare it with the plug used in the next test. Otherwise the differences caused by changes within the fine tuning range will be too small to be noticeable, or, they will lead you to make greater-than-necessary changes to the carburation, etc.

HT leads

The leads and plug caps must be kept in perfect condition – and renewed if there is any doubt; a faulty lead can create spurious effects. Leads should also be kept as short as possible and not run parallel with either one another, with any other wiring, or with a frame tube, otherwise there is a risk that the rapidly-changing current will induce a voltage in the neighbouring wire, giving unpredictable results.

Coils

The stock items will probably not be suitable for an engine which is tuned by more than 15 per cent above its original output. High-performance coils or car coils are available from specialist suppliers and they should be mounted where they will receive some cooling air flow, without subjecting the leads to too much rain or road spray.

Many coils do not build up current quickly enough to be able to work efficiently at high speeds. It is possible to use a coil of a lower rating than the electrical system on the bike, in conjunction with a heat-sensitive ballast resistor. When the coil's current consumption is correct, the heat build-up in the resistor increases its resistance and restricts the voltage applied to the coil. At high speeds, when the current falls, the resistor cools, allowing a greater voltage to reach the coil.

Amplifier

This takes the signal from the trigger/sensor unit and executes the capacitive or inductive discharge in the coil's primary circuit. It may include an electronic advance mechanism and a rev limiter, which cuts out some or all cycles at a pre-set frequency. It obviously has to be tailored to the rest of the ignition circuit and to the engine's requirements. As there is no provision for adjustment, a firm specialising in competition ignition systems should be consulted.

Trigger
This has to match the amplifier and has to be capable of fitting the engine; the choice of amplifier will probably dictate the type of trigger used.

Timing
The most effective method is to use a dial gauge mounted vertically to follow the piston travel, and to set the timing point in mm BTDC, marking the rotor etc., in this position, so that a stroboscope can be used when the engine is running. Some triggers do not have any facility for static timing

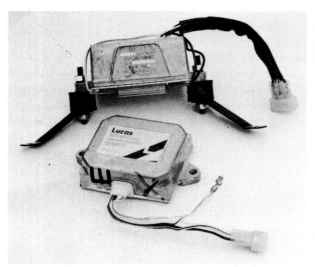

Fig. 52. Lucas RITA replacement amplifier units can be tailored to a limited extent to suit individual applications

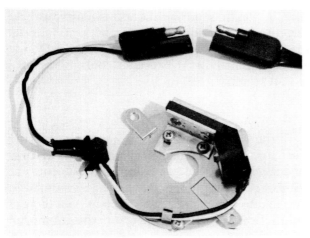

Fig. 53. Replacement trigger unit

and must be set up approximately so that the engine can be run and the final timing set by using a strobe.

Once the correct static position has been established, future work can be speeded up by holding the crankshaft still while fitting the cylinder head and then modifying an old spark plug to take a plunger, which is positioned to form a stop for the piston. The easiest way is to break the ceramic insulator from the plug and then weld a nut on to the metal body so that a long screw can be threaded through the plug to reach the piston.

Minimum advance

The optimum ignition timing is called MBT or minimum advance for best torque. It is quite clearly seen in dyno tests where you can hold the motor at a steady speed and progressively move the timing through several degrees. When the timing is retarded beyond the optimum, the load drops noticeably. Advancing past the MBT point usually gives small increases in load but they will be accompanied by higher combustion temperatures and the engine will be more prone to detonation. The dyno test should identify this point, where the graph turns a sudden corner, and the optimum ignition timing – at that speed – is the minimum amount of advance to keep the load on the high side of the turning point.

It is quite likely that the MBT point will vary at different engine speeds. If this is a fairly steady progression then the advance curves built in to many ignition systems will approximate to it. Get it exactly right at peak torque, where detonation is most likely to start, and let it be approximately right everywhere else. Systems with digital control can move the timing position to wherever it's needed – there are a few specialists who can modify existing systems or who can supply programmable replacements to cope with modified engines.

Chapter 9

Lubrication and cooling

Most four-stroke roadster engines have wet sump lubrication with a pressure feed to the gearbox. There are several ways in which this sytem can be altered, when required, to assist an engine modified for competition use.

While the oil is there to lubricate, it also cools the the engine and many designs use the oil feed as an additional, local coolant. Typical examples are the use of an oil spray directed on to the underside of the piston, or an oil gallery routed to pass close to the exhaust valve seats. To permit a high flow rate where it is needed without over-oiling or building up excessive pressure in other places, restrictor jets are often placed in the oilways, usually in the feed to the gearbox and sometimes in the feed to the valve gear. Two layouts are shown in Fig. 54, including a variation which has two circuits, one going simply through the oil cooler and back to the sump, the other circuit (in parallel with the first) taking the usual route through the engine components.

Various lubrication problems can occur when an engine is modified, or sometimes when it is simply used for long periods at high speed. On the other hand, some lubrication modifications may improve the performance or the reliability of the engine.

Problems

1. Blocked oil jets or oilways, especially after the casings have been bead blasted or degreased in an ultra-sonic tank. Before rebuilding, degrease and wash all oilways with a low-pressure water jet, flowing in the reverse direction. Blow the oilways dry with an air line. Make sure all of the restrictor jets are in place, as some only rest in position, being trapped by the crankcase halves on assembly.
2. Leaks from stud holes used as a return passage for the oil – usually where O-rings or gaskets do not have enough sealing area or the engine block has allowed distortion after a big-bore modification. Use a better gasket or one which is compatible with the new head gasket.
3. Leaks from a bored block – usually because the surface has not been sealed. Shot peening before the liner is fitted should help.
4. Oil swill under acceleration or braking. This can be controlled by making up sheet steel baffles to fit into the sump (a) to prevent the level swilling away from the pick-up pipe (although this is usually catered for in the stock design) and (b) to prevent the oil level rising and reaching

Fig. 54. (a) Arrangement of the wet sump system used by Suzuki on the GSX-R750, showing how the oil is used to cool the pistons and cylinder head. (b) Dual loop system used by Kawasaki

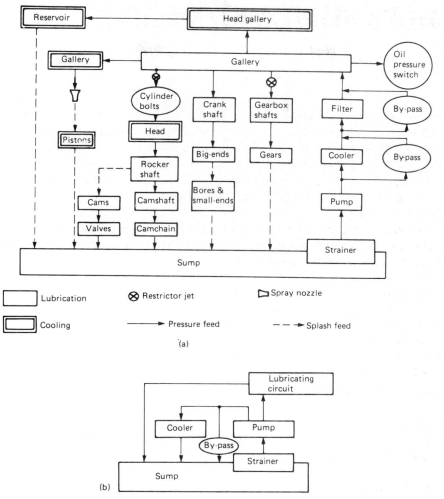

the crankshaft, which will cause drag and the excessive splashing may cause an oil control problem at the piston rings.

5. Oil control – where crank journal tolerances are run on the upper limit or the side clearance is increased, where the engine speed is raised or two-ring pistons are used (see Chapter 4).
6. Excess pressure. This can damage the matrix of the oil cooler – a pressure relief valve should be used – and can damage oil seals or filters. A pressure relief valve is fitted as standard but it can be overloaded if the pump is uprated or if the engine is run at high speed with a heavy oil/cold oil. Do not uprate the oil system without providing somewhere for the oil to go.

Fig. 55. Dry sump kit fitted to an in-line Honda RCB

7. No breather oil trap when the air box is removed. Some K & N individual filters have provision for an engine breather connection, or a separate catch tank must be used. In either case an oil trap — a box containing baffles designed to catch oil droplets and let them drain back into the sump — should be fitted.
8. Revised cam or rocker geometry may shield the working surfaces from an oil spray or splash feed. If so, the drilling will have to be changed or opened up to compensate.
9. Increased valve spring force raising the contact stress at the cam/follower. An oil with better boundary lubrication properties is the only way the lubrication system can deal with this problem. It may be better to ease the mechanical stress by using lighter springs or grinding a smaller radius on the follower.
10. High oil consumption; breather fills air box with oil. Several factors can influence this — which is associated with continuous high-speed running. First, better oil control rings or rings less prone to flutter (see Chapter 4). Second, a heavier oil or a type less prone to boil off. Third, a greater breather capacity with a more efficient oil trap. Fourth, improved high-speed cooling, or a dry sump kit.

Modifications

1. High delivery oil pump (or high ratio gear drive). Should only be used if there is somewhere for the extra oil to go, e.g. greater crank clearances or larger oil jets to provide extra piston cooling, which may be necessary on supercharged engines.

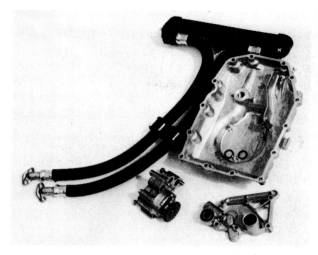

Fig. 56. In some cases, the lubrication system can be uprated by using components from a larger model – typically the sump pan and oil cooler connections, plus the oil pump

Fig. 57. Power fade. The output of an air-cooled GPz550 Kawasaki running at peak torque (8500 rev/min) while the temperature was varied by restricting the cooling air flow. The temperature was measured by using a washer under one spark plug as the hot junction of a thermocouple.

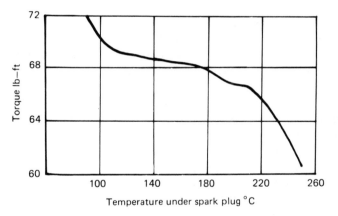

2. Oil cooler, or a larger oil cooler. Sometimes it is possible to use the connections or the complete sump pan from a larger model. Always use a pressure-controlled by-pass.
3. Baffles to fit crankcase (see above) plus a scraper – a screen kept as close as possible to the crankshaft webs to prevent oil being carried around with them.
4. Greater breather capacity. A larger crankcase breather, plus additional breathers from the gearbox and cam boxes, used in conjunction with oil traps and catch tanks.
5. Larger jets or drillings to increase oil spray to underside of pistons or to camshafts. A similar cooling effect can be made by running the big-end

journals at a larger clearance (within tolerance) or by using more end-float. Better piston cooling may also be accompanied by oil control problems.
6. Dry sump conversion. Reduces engine height, gives better control over bulk oil temperature to prevent the lighter fractions boiling off, and avoids the problems of oil swill in the sump. The kit needs internal baffles, high capacity scavenge pump(s), external oil lines and a separate tank.
7. Increased oil flow to cylinder head – to improve cooling or to increase supply to new cams – achieved by uprating pump, using smaller restrictors elsewhere or larger restrictors in the feed to the head.
8. Using engine clearances to control oil flow – e.g. bronze alloy valve guides run tighter clearances, as long as they are compatible with the valve material (see Chapter 5), and give better oil control.
9. Dry clutch or uprated clutch (see Chapter 10).
10. Oil cooler thermostat, typically opens at 75°C.

The type of oil to be used must be considered in conjunction with any proposed modifications. The best load-bearing lubricants are the heavier viscosity oils or the EP types but both of these do not flow as well as light oils and tend to build up heavier deposits, so their cooling properties are not so good as lighter oils. Vegetable oils have the best lubricity and give the best protection in borderline conditions but tend to form heavy varnishes and, as they are hygroscopic, they deteriorate fairly rapidly. Synthetic oils possibly make the best compromise for highly-stressed engines; the oil companies' competitition departments are usually very helpful with recommendations for specific applications. Where heavy oil is used, it is essential to bring the engine up to full working temperature before using it under heavy load or at high speed.

Air cooling

Although the thermal efficiency increases at higher engine temperatures, mechanical efficiency decreases and there have been many tests which show a marked drop of power as the temperature is increased. As materials technology progresses it becomes possible to run engines at higher temperatures without mechanical losses and therefore possible to benefit from the improved thermal efficiency and this is where the use of ceramics – in pistons, liners and valves – is likely to have its biggest effect (see Chapter 4).

But where aluminium alloys and steels are used together in pistons, liners, heads and valves, the operating temperature will continue to have a dominant effect upon performance. On many air-cooled engines the output falls noticeably once the cylinder head temperature (measured under the spark plug) reaches a critical point. This varies from engine to engine but is usually in the region of 200 to 250°C.

Improvements need to be concentrated around the upper cylinder and head. Ducts are an obvious answer, particularly where a fairing is used. Turbulent air flow, as long as it is directed to the right places, is the most efficient way to cool the engine for the minimum amount of aerodynamic drag, so an entry behind the forks or mudguard is not necessarily a bad thing. There must also be an exit, to lead the air flow away from the engine and it is important not to flow warm air over the ignition coils, carburettors or their air intakes. The same applies to any ducting used for oil coolers.

The results of any tests should be checked using a thermocouple with the hot junction connected to a washer so that it can be fitted under a spark plug or some other convenient nut on the engine.

A before-and-after sequence of tests should be repeated many times because the ambient temperature and the wind direction will have a more pronounced effect than the ducting.

Drilling holes in the finning will have several effects, depending on the diameter of holes and the thickness of the fin. As well as reducing the amount of material and making the engine lighter, a hole will alter the surface area as follows:

for $t > d/2$ surface area will increase
 $t = d/2$ surface area will not change
 $t < d/2$ surface area will decrease
where t = thickness of metal
 d = diameter of hole

The last two cases are likely to make the component run at a higher temperature as there will be less surface area and less mass, so a given heat input will create a higher temperature. In the first case, where the radius of the hole is less than the thickness of the metal, there will be an increase in surface area which will potentially give better cooling. This particularly applies if the air speed over the fins is low or turbulent, but it will be offset to some extent by the loss of material.

Liquid cooling

This system makes it easy to apply cooling precisely where it is required and to control it – thermostatically, by positioning the heat exchanger in the air flow or by masking part of the heat exchanger.

Deposits and corrosion in the coolant passages or on engine surfaces will insulate the engine and obstruct heat flow, so do not use additives (especially radiator sealant) or hard water. Use soft or distilled water, with an anti-freeze which contains an alloy corrosion inhibitor.

Gaskets must not obstruct any of the passages – except where they are designed to form a baffle. It may be possible to increase the pressure of the coolant locally (usually near the exhaust valve) by fitting a V-shaped restrictor in the outlet from this region. This may prevent local boiling and the formation of a vapour pocket. The engine thermostat (if fitted close to the exhaust side of the head) will perform the same role as well as controlling the coolant to give the shortest warm-up time. The optimum setting will be between 75 and 90°C – the ideal temperature for a particular engine can only be found by experiment.

To bleed the cooling system, remove the radiator cap and, having filled the system and checked for leaks, run the engine until it is warm. Check the level of coolant and, if necessary, top up the header tank or reservoir. Make sure that the bleed tube from the reservoir to the radiator is connected and that none of the hoses are kinked.

Chapter 10

Engine preparation

The way in which the engine is built is critical to reliability and it also has a measurable influence on power production. Power is the sum of the heat released inside the cylinder minus the losses due to friction, oil drag, pumping losses, crankcase losses, and the power taken by ancillary equipment. Good engine preparation will reduce some of these losses and consequently raise the available power.

Where blueprinting is concerned the power gain is worth having. The difference between a brand new engine and the same engine after running-in, can be in the region of 10 per cent; blueprinting is nothing more than ensuring that the clearances are at their optimum settings, so it can be expected to make a similar sort of difference.

In a modified engine there are two aspects to preparation. One is meticulous assembly, building the motor exactly as it says in the shop manual. The second is the modification of parts to withstand higher stresses or to suit the new conditions. So a set of competition regulations and a shop manual need to be added to your own notebook.

It may be necessary to exchange some parts for better quality items; it will be possible to modify other parts; in either case it is necessary to have some idea of their ability to withstand stress and of their fatigue resistance (see Appendix).

This assumes that the engine builder has the knowledge, skill and the equipment to do the job in the first place and the realization that there are big differences between a tuner and a mechanic; mainly that getting it right has precedence over how long it takes.

A look at the various steps of an engine build will show what is required:

Gearbox
The oil supply is often metered through jets, which must not be obstructed. The jet size may need to be altered if the pump is uprated or the oil system modified. If any changes are made to the transmission, and particularly to its lubrication, then they should be followed by a series of careful tests to make sure that the transmission is working reliably.

Close ratio gears are available for several models, particularly where the engine is suitable for a racing class such as F1 or F2; often these are available from the manufacturer, but they are expensive.

Fig. 58. Close ratio gear cluster, available as part of a Kawasaki race kit

For special applications e.g. drag racing, there may be purpose-built gear clusters, sometimes containing only two or three gears and possibly using an air-shift control. American specialist suppliers are the best source of this type of equipment.

Often the stock gears will not take the abuse they get under racing conditions and the gears may jump out of mesh. In this case it will be necessary to have the dogs undercut (maching a slight taper on the loaded

Fig. 59. Competition clutches or dry clutches (such as this RSC race kit) are sometimes available from the manufacturer or from performance shops

Fig. 60. The alternator on this Kawasaki has been replaced by a toothed-belt drive to an ignition distributor

face of the dog) which, because of the accuracy of the work and because of the hardness of the material, requires special equipment.

Clutch

This can be progressively uprated to keep up with the output of the engine. Stronger springs, or a washer under each spring, is the first step, followed by the use of competition plates. Distortion of the clutch basket is the next problem, and exchange units are available for certain models with stronger shock-absorber springs fitted, or with the basket machined from a solid billet. Heavy-duty thrust bearings are also available from performance shops.

If the frame layout is cramped, it may be worthwhile using an hydraulic clutch lifter, by adapting one of the hydraulic systems being used on stock models or by modifying a rear brake master cylinder to work as a slave cylinder and move the operating linkage. This saves the need to route a heavy-duty cable smoothly around engine and frame, and it also takes up adjustment automatically as the plates get hot, an advantage in sports like drag racing where a lot of full power starts have to be made.

Roller bearings

Heavy-duty bearings are available from performance shops for certain models. Alternatively, if the stock bearing code numbers are known, a bearing supplier will be able to check their specification and may be able to offer an uprated version.

Fig. 61. This endurance racer has also had its alternator relocated above the gearbox, this time with a V-belt drive from the end of the crankshaft

Transmission

Gear clusters can be modified in several ways. It may be possible to change some or all of the ratios, from a factory race kit or from specialist suppliers, to get ratios which are more suitable for racing conditions. Usually this means a higher first gear and smaller gaps between the ratios to allow for an engine with a narrower power band. Plotting the road thrust curves or using a program like RL (see Appendix) will show whether this is desirable.

New gears can be made which will shift faster (usually by having three instead of six dogs) and will be less likely to jump out (by modifying the slope of the undercut ramp on the driving face of each dog). Gearboxes for drag racing often feature long ramps on the reverse face, which make the gears slide together during shifts, with the minimum disturbance to drive. It also means that this pair of gears cannot accept reverse thrust (i.e. engine overrun) because this would make them disengage, possibly damaging the selector fork, pushing the shaft through its bearing or selecting another gear (which could destroy the engine/crankcases).

If there is a problem with gear reliability (usually teeth breaking at the root, sometimes a surface fatigue failure on the face of the tooth) then stronger gears can be made using a new tooth profile or wider wheels (which have to be matched to the mating gear) or stronger materials/heat treatment.

There are a few specialists who can supply this kind of service but because it is so expensive to make one-off gear clusters, they usually concentrate on two or three popular machines. Other modifications include billet clutch baskets which are stronger than the original equipment and straight-cut (instead of helical) primary drive gears. While helical teeth give smoother drive and less noise, they create side-thrust which is all wasted effort. If helical gears are used, they must be shimmed to give the specified amount of backlash in order to keep frictional losses to a minimum.

Crankcase

Dry sump conversions are available for a few models; on others there may be optional sump pans (often from a different model in the same range) which may offer some advantage, like a change in capacity or oil cooler connections.

The oil pump clearances, seals and drive should be checked carefully. It may be possible to uprate the pump, either with a new body or by changing its drive, although the reasons for wanting to increase the oil flow should be examined carefully.

Steel baffles should be made up to bolt into the lower crankcase to prevent oil swill (see Chapter 9).

Security wire should be used to lock oil drain plugs and oil filter housings before the engine is used on a circuit.

The crankcases are subject to the same fatigue loads as the crankshaft and it may be necessary to stress-relieve the inner casings, particularly where casting flashes or bosses have been left near the bearing housings.

Check the depth of all stud holes and stress relieve (see Appendix) and use high-tensile studs, particularly on the crankcase/barrel joint.

The truth of the main bearings should be checked – in a line-bore jig – and if the crank is offset to the cylinder bores then the thrust forces on the side of the pistons will be increased.

Crankshaft out of balance forces are transmitted through the main bearings, so the housing and crankcase walls have to be strong enough to take these forces. Where balance shafts are fitted, the forces are contained between the two shafts, allowing the rest of the casing, engine mounts etc. to be much lighter. However, the webs supporting the bearings are subject to these large fatigue forces and if the engine speed is increased or the out of balance forces are increased by using heavier pistons, then this may exceed the material's endurance limit.

This can be improved by grinding away any stress raisers (see Appendix) in the form of casting flash, seams, any irregularities or sudden changes of section, which should be radiused as generously as possible. The surface should then be polished and shot peened to toughen the surface layer and close up any minute cracks. It is essential that the shot peening process is one designed for this purpose and not the type used for cleaning components.

Crankcase breather

The capacity of the breather should be increased by as much as is physically possible. Although four-cylinder engines have no overall crankcase displacement, there is nevertheless a lot of gas movement. The breather helps to provide a surge tank to absorb this movement, as well as coping with any blow-by caused by ring flutter. A lot of oil can be lost via the breather at high crank speeds and the larger its capacity, the lower the gas speed in it. It should have an oil trap, a de-aeration chamber and a return to the sump – possibly adapted from the original fittings in the base of the air box.

Crankcase bolts

Most of the bolts around the engine can be replaced by better quality items, particularly the screws holding the crankcase covers. While this is not essential, it is a big help if the engine is being stripped and rebuilt frequently.

Mating castings must not be located by the bolts which hold them together; this job should be done by dowels. Most Japanese engines have enough dowels but if not, more can be bored, as long as the castings can be set up accurately using the crankshaft bearings as a register. If there is not room, a hollow dowel can be fitted over one of the studs. Some roller bearings are also dowelled or pegged to prevent them spinning in the outer housing.

Fig. 62. Balance shafts may need to be modified to suit heavier pistons

Some crankcases distort if there is an alignment problem and when they are tightened up they pinch on the bearings, making it difficult to turn the crankshaft by hand. In this case the problem must be located and the crankcase halves modified to align accurately before the engine is built.

Thread lock
Use the recommended grade of locking agent on moving internal parts like the gear selector mechanism.

Balance shafts
All engine inertia forces will be contained between the balance shaft and the crankshaft, so it is not a good idea to remove it. The casings and engine mounts may not be strong enough to take the newly-liberated forces.

If heavier pistons are used, then the balance factor should be altered accordingly (see Appendix) by drilling or machining the side opposite the counterweight. Stress relieve the shaft and its bearing housings if necessary and carefully tension the drive chain.

Crankshaft
It is often necessary to shorten the crankshaft, partly to reduce engine width and partly to give it a higher natural frequency, in order to avoid torsional vibration (especially if the crankshaft is lightened and used at higher speeds). If the alternator is carried on the end of the shaft it should be removed or replaced by a lighter competition unit. For endurance racing, the alternator can be relocated above the gearbox, driven from the crank or from

a pulley on the gearbox sprocket. A blanking plate with a suitable gasket will need to be fitted over the end of the final oil seal. The crankshaft may also have to be modified to accept new ignition pick-ups although sometimes these can be located on the camshaft.

The crank can be lightened by machining the webs, at the same time all bearing shoulders should be given the largest possible radii. Polishing and shot-peening will also help to remove any surface stress-raisers, while a crack-detecting test is the only way to be sure the unit is not flawed. If the crank is lightened, the balance factor to suit the piston mass should be kept.
One-piece cranks: measure the shell bearing clearance with Plastigage and select shells to give the required setting. It may be advantageous to run the maximum permissible tolerance at the big-end journals, although this will increase the oil flow through each bearing and may overload the oil control rings.
Shell bearings: there are several types of shell bearing, with materials selected for different applications: White metal – not able to handle the high loading involved. Aluminium-tin – more common in motorcycles but is sensitive to small reductions in lubrication and picks up immediately. Lead-bronze – needs to be used in conjunction with hardened journals, good load-carrying ability and less sensitive to fluctuations in lubrication.

These are sometimes called tri-metal shells, as they are made by having a steel backing with the composite metal overlay either cast or sintered. On a lead-bronze shell bearing, the composite overlay is 0.0127 to 0.0343 mm thick.

The recommended load capacity of shell bearing materials is:

cast leaded-bronze	7,000 lb/in^2
aluminium-tin	4,500 lb/in^2
lead-based white metal	2,250 lb/in^2
tin-based white metal	2,250 lb/in^2

Crankshaft runout: to check the crank's runout, set the outer main journals in V blocks and rotate the crankshaft, using a dial gauge on the centre main

Fig. 63. This endurance racer has the shortened crank blanked off at the main-bearing oil seal and carries its alternator on a plate, with a belt drive running from the gearbox sprocket

Fig. 64. A dial guage set up to measure crankshaft end-float

journals. While a typical service limit is 0.05 mm, an acceptable limit for a competition crank is 0.01 mm.

Built-up cranks: the advantages of roller bearings (less friction, less sensitive to lubrication) and the better rod design are offset by the inherent weakness of a pressed-together crankshaft. However, the crank can be overhauled and rebuilt with heavy-duty bearings, although this is a job for a specialist. On race engines, particularly for drag bikes, the crankpin should be welded to the web to prevent the crank twisting.

When the engine is built, the accuracy of the crank and the cylinder block should be checked to ensure that all pistons arrive at TDC in multiples of 180 degrees (or the V angle in the case of V-motors).

Connecting rods

These are usually made from either forged steel, aluminium alloy or titanium. Grind off the forging flash on the seams and make sure that the big-end bolts are a very tight fit in their holes, while all recesses have radiussed inside corners. There must be no notch or scratch marks on the rod.

The big-end bolts should only be used once and the bolt loading should be checked by measuring its stretch rather than the tightening torque, although the shape of the bolt head often makes this difficult. If the stretch figure is not available it can be obtained by taking a new, clean, dry bolt and tightening it carefully to the recommended torque, measuring the length before and after with a micrometer. The same stretch figure should then be applied to all the other bolts. Torque wrench settings are usually applied to dry, solvent-cleaned threads – but check with the shop manual first.

Rods should be checked for bend and twist; the stock limit of OE rods in both cases is 0.2 mm/100 mm. Rods should also be selected for equal weight (often they are weight coded by the manufacturer but will only be matched in pairs).

Fig. 65(a). Competition connecting rod; stress-relieved, polished and shot-peened

Fig. 65(b). Alternative competition rods may be available from specialist suppliers. This one is made by Carillo

While the big-end bearing clearance has a major effect on the flow of oil past the journal, the big-end side clearance also has some control over the flow and the resulting pressure drop. This flow is responsible for splash lubrication on to the cylinder wall and the underside of the piston. It therefore needs to be regulated quite carefully. Centrifugal force tends to increase the flow to the big end, a force which will increase if the crank speed is raised. The stock figure for big-end side clearance is 0.10 to 0.20 mm on high-speed engines, with a service limit of 0.3 to 0.4 mm; the stock figure can be enlarged slightly in order to increase the oil flow rate.

Competition rods are available for some engines and should be used if the piston mass or the peak crank speed has been raised.

Finally, longer or shorter rods may be available from other engines with the same size journals or from a race kit, for example Kawasaki supplied longer rods for their GPX750. A longer rod will give lower piston acceleration and cause less side thrust, while a shorter rod can give better low-speed torque. The calculations for the precise effects are given in the Appendix.

Fig. 66(a). Race pistons often have the non-thrust part of the skirt cut away and use two rings. The tapered part of the skirt just below the second ring sometimes has oil drain holes drilled in it

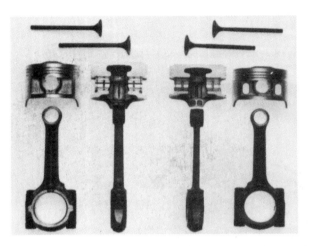

Fig. 66(b). Development of Honda's V4; compare the reciprocating parts of the VF750F (*left*) with those of the later VFR750F (*right*). The thinner rods are 90 g lighter, strength being maintained by carburizing the surface; the pistons are 20 g lighter, with thinner rings; the valves are waisted, saving 0.5 g per valve on the intake and 1.5 g on the exhaust. In addition the rockers were made 6 g lighter and the valve springs lightened by 17 g per valve (on a 24-valve engine).

Piston

The high silicon content, cast pistons used by Japanese OEM are very good; they are nearly as strong as forged items, are lighter and resistant to scuffing. Stock pistons have proved themselves to be good for 125 bhp/litre and should be used where the emphasis is on high crank speed rather than high thermal loading. If the heat flow is increased significantly, for example if the engine is turbocharged, then forged pistons will be needed.

Solid-skirt pistons should be used, the bore being honed to give the required piston-cylinder clearance and oil control being handled by this, the bottom ring and relief holes drilled in the bottom ring groove or just below the bottom ring.

The piston should be measured 5 to 10 mm below the piston pin and at right angles to it; manufacturers usually quote a distance up from the bottom of the piston skirt – keep carefully to this method. The cylinder bore should be measured in six or eight places (see below) and the piston clearance is the bore size minus the piston size. This will vary in the region 0.04 to 0.07 mm on stock engines, with bore sizes around 60 mm and the stock clearance can be opened up by 0.010 to 0.015 mm on modified engines, using cast pistons; forged pistons may require slightly more clearance, according to the recommendation of the piston manufacturer.

If there are problems caused either by the crown expanding too much or by piston rock (both allowing the ring lands to bear on the cylinder) then the

Fig. 67(a). Measure pistons ar right angles to the piston pin and in the same position on the skirt, usually just below the pin

Fig. 67(b). Piston proportions. The intricacies of shape are shown by exaggerating the horizontal scale in these two drawings, showing (*left*) a Cosworth piston and (*right*) a piston made by Cagiva

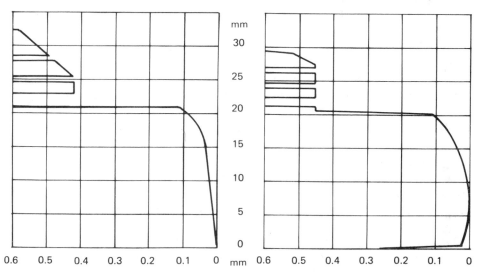

piston can be re-proportioned by machining the ring lands down by 0.02 to 0.05 mm, as necessary.

The pistons should be matched closely for weight and deck height.

Piston rings should fit the grooves perfectly – if a groove is damaged or out of parallel, the piston should be rejected. Check the ring groove clearance in several pistons; the figure will vary according to the type of ring but, typically it will be 0.03 to 0.06 mm for a plain ring, and slightly less for tapered, etc, types used as second rings.

Piston pin and small end

The most neglected bearing in the entire engine, the small end often shows signs of overheating on standard engines. Despite this, it does not appear to suffer any adverse effects when the engine is modified, other than shortening the life of the piston/rod. As these items rarely see a full service life anyway, the neglect of the small end is not a critical factor. On some engines it may become critical and then it will be necessary to improve cooling/lubrication, or to find a rod with a larger bearing area.

Heavy circlips, while easier to fit than the bent wire type, are not necessarily better. Under the immense inertia loading at TDC, the circlip can lift out of its groove – and may not return into it. Instead, the circlip finds its way out of the piston and then it or the piston pin proceeds to destroy the cylinder. This suggests that the circlip gap should be facing the piston skirt. The alternatives to circlips are PTFE pads, available from American performance specialists, or spire-locks – circlips used with machined pins which force the circlips into their grooves.

Cylinder

Test the truth of the block by checking that pistons reach TDC at the angular intervals set by the crankpins and that the top and bottom faces are flat and parallel. If the error is in one or other faces it may be possible to machine it true and to use a bottom gasket thick enough to make up the difference. If the bores are offset or not parallel then the block should not be used.

The bore should be measured along the piston pin axis and at right angles to it, in three or four positions, just below TDC, and at intermediate positions down to the unused, bottom portion of the liner.

The limits for taper/ovality are less than 0.01 mm between any two measurements.

The bore should be honed to adjust the piston skirt clearance and to give a good, cross-hatched finish which will retain oil and help the rings bed in quickly. The cross hatch angle should be about 60 degrees, with a plateau area of 1/2 to 2/3.

Where plated bores are used, the piston has to be selected to give the required clearance.

Cylinder studs

These are subjected to cyclic tensile loads each time the engine fires and any stretch, or deformation at the threads is a potential source of head gasket failure, while the studs or the casings may suffer fatigue failure.

Fig. 68. Measure the bore with an internal micrometer or a bore gauge in several positions along the length of the bore, in line with the pin and at right angles to it

Fig. 69. Machining grooves in a cylinder head to accept inert gas-filled rings

High-tensile studs must be fitted, preferably with rolled, not cut, threads and the studs and holes stress-relieved (see Appendix). Dowels must be used to locate the block and the cylinder head and, if cap head screws are used, the bearing area under their heads can be increased by using flat air frame washers.

133

Stretch in the cylinder and head studs occurs when they are tightened, when the engine expands thermally and each time the engine fires. The studs behave like springs, whose pre-load is the initial tightening torque. If the initial tensile load is fairly low, then the forces due to expansion and combustion will be large compared to it and the fluctuation between minimum and maximum will be at its greatest, that is, the fluctuation in the clamping force at the head gasket will also be at its greatest. A 'soft' gasket material, provided that it isn't deformed beyond its elastic limit, is meant to accommodate this fluctuation but it may take the clamping force below the level necessary to seal (particularly if tuning has raised the combustion pressure which will increase both the stretch in the studs and the sealing requirement) or it may cause fatigue failure at the gasket.

The use of rigid head designs and hard gaskets (see below) make this more critical. In this case the studs are tightened as much as possible, to make their pre-load as high as possible in relation to the maximum load they will receive and therefore to keep the fluctuations in strain to a minimum. As the tightening force needs to exceed the clamping force at the gasket but doesn't need to be much greater, the studs should be sized so that they provide this force when they are tightened close to the material's yield point. (Some engine manufacturers design the studs to be tightened just beyond the yield point, so expansion when the engine first gets hot deforms them even more, causing strain hardening of the stud material. This toughening process leaves the steel below its new elastic limit after the engine cools down; because it now has a higher elastic limit, subsequent engine running keeps within the new elastic range of the studs. The result is that there is ample clamping pressure to avoid gasket failure whether the engine is hot or cold – and can therefore be run under load without warming up – and there is no need to retighten the studs at all once the initial build is finished. New studs must be used each time the engine is rebuilt.)

Tightening torque – as screw threads are tightened, more and more force at the spanner is needed to overcome friction in the thread and so less is available as tensile force in the stud or bolt. Consequently torque wrench readings are less reliable (in terms of the clamping force being generated) at higher torque values. It is usually not possible to measure bolt stretch, which is the only accurate method, but the next best is to specify a low-ish torque setting, where thread friction is not too significant, and then to tighten the nut a further, specified distance. Torque wrenches are available with a degree scale for this purpose.

Head gaskets

On well-prepared surfaces, stock gaskets are perfectly acceptable.

There are various alternatives to the stock head gasket, such as plain steel, copper, alloy or graphite (which allow better heat transfer from the head to the cylinder) and various pressure-backed types or gas-filled rings, some of which may need grooves machining in the gasket face. Gaskets exposed to the combustion chamber may suffer fatigue failure.

Current thinking tends towards the use of a thin, hard (e.g. steel) head gasket. The reasoning is this: the head and gasket are squashed in compression, while the studs are stretched; expansion when the materials get hot increases the compressive force on the head; gas pressure during combustion increases the tensile force on the studs making them stretch and easing the clamping force at the gasket. If the gasket (or the head itself) is easily compressible then the initial clamping may squash it beyond its elastic limit and it will stay deformed. Further deformation caused by thermal expansion will be permanent and the clamping force will be reduced when the engine cools down; it will also be reduced during combustion, possibly to the point where the gasket fails to seal or, after a certain amount of running, suffers fatigue failure.

This is why traditional, soft, gasket materials (e.g. copper) must be clamped to a specified load and not overtightened; must be re-clamped after the engine has been brought up to running temperature and cooled down; must not be re-used.

The alternative, to make the head and its gasket as stiff as possible, relies on the fact that this will reduce compressive strain and keep it well within the elastic range of the materials. The tensile force in the studs then merely needs to be sufficient to clamp the gasket beyond its sealing point at high temperature/full load conditions. Overtightening should not be a problem (until it stretches the studs or threads beyond their yield point or unless it is uneven and therefore distorts the assembly) and retightening after the initial engine run should not be necessary. The gasket could be re-usable.

Cylinder head

Preparation here includes a thorough cleaning and a check that all of the oilways are clear and align with the block/head gasket. The head gasket face must be perfectly flat – checked with a straight edge, or set up in a mill, which can be used to skim the surplus from a warped head.

Where the head surface of the combustion chamber does not exactly align with the cylinder, or where there is a squish band, or where the valves have been set back into the head, there must be no sharp edges. Put a gentle radius on all corners, to reduce the chance of pre-ignition. All combustion chambers should have the same volume.

If the head surface has been machined, then it will be necessary to alter any head steady, top engine mount or external oil feed which is fitted.

Check all the valves and springs for the correct fitted length after lapping the valve seats, and replace any springs which have taken a permanent 'set'. Carefully examine all of the valve gear, including the locating collets and

collars, for stress raisers, and stress relieve if necessary (see Appendix).
Check that all intake and exhaust stubs align, with no steps or ridges.

Cam drive

Where chains are used on high-speed engines, their weight and quality are the main criteria. A heavier chain will not be a better one. Automatic chain tensioners have been known to back off under extreme conditions and it is common practice to modify the tensioner so that it has to be adjusted manually and then locked using a bolt and locknut. Another type of tensioner can sometimes be adapted, for instance that used on the Kawasaki Z250, which will fit the larger Kawasaki engines (see Fig. 31).

Where a shaft or gear drive is available, adjust the backlash carefully by shimming the gear wheels and checking it with a dial gauge.

Cams

Having set the timing, check it on all valves (see Chapter 5) and check for any eccentricity at the cam drive sprocket – which will cause instability in the chain drive. There should be no measurable runout at the sprocket.

Cam box

Add a breather, if one is not fitted, by boring a 10–15 mm hole and welding a spigotted tube into it; lead the fumes into the main breather catch tank.

Valve clearance

Where bucket and shim adjusters are used, the type which has the shim on top of the bucket, while easy to adjust, is prone to failure at high speed because the shim can fall out of position if the valve floats. For this reason it should be changed for the type which has the shim located under the bucket.

General assembly

Build the engine dry for the purpose of taking measurements, etc. On final assembly, do not lubricate threads unless the shop manual specifically requires it; lubricate bearing surfaces with engine oil if the engine is to be started immediately after assembly. Where it is to stand for some time, lubricate bearing surfaces with an anti-seize assembly compound. Before running, the engine should be motored to circulate the oil, and to check that oil is being fed to the furthest components such as the valve gear.

Chapter 11
Fine tuning and testing

It is preferable to develop an engine in single steps, so that only one thing is changed at any one time and the effect (and cause) of each change can be clearly seen. In practice this is rarely possible, partly because some changes *need* to be made in conjunction with others before they can work properly – and partly because there is not time to go through all of the carburation and ignition settings to suit each step.

The result is that the carburation and other settings are not set precisely until the very end – and this sometimes means that it is very difficult to make any sense out of the engine at all. Bearing this in mind, it is a good idea to work out a sequence of steps and tests which involves the minimum of rebuild/tuning time an still manages to show a progressive increase in performance. The logical way is to make the biggest changes first, and finish with the detail work.

A typical programme might start with a big-bore conversion, followed by a test run which should indicate a lack of valve time-area, and possibly a mismatch between engine and carburettor. The bigger displacement would increase the load, but the mismatch between capacity and valving would prevent the engine performing properly at high speed.

The next step would be to modify the cams to give the required valve area, but to speed up the work, this operation would also include any work done to the cylinder head, valve size and so on, unless it was necessary to find out how much effect a particular modification made. The next test would have to include experiments with carburettor and intake dimensions and with changes in the exhaust. It would probably be easiest to establish the length of plain exhaust pipe which gave the best peak output and to leave this fitted while the best intake length and carburettor size was established. It would obviously be necessary to tune the carburation, but only to the minimum extent necessary to conduct the tests; that is, WOT running with a mixture strength that erred towards the rich side.

If no problems occur up to this point, the exhaust system is the next item to be developed – this time in some detail and with carburettor adjustments to suit each step. From here onwards, the work falls under the broad heading of 'fine tuning', simply because each step will only give a small increase in power and getting the correct carburation and ignition timing will be worth a similar amount of power.

Once the engine is roughly in its final form, the compression should be increased until it reaches the practical limit imposed either by the fuel or by the mechanical clearances.

The ignition timing will have to be adjusted to get the best torque without knock and it may be necessary to experiment with different coils and different grades of spark plugs. At this stage, it is quite likely that problems will occur. Something may have reached its limit and will have to be modified or replaced with a tougher part – or it will be necessary to go back to the level at which the engine is reliable. It is also possible that some modifications were not extreme enough – the engine may need more cam duration to reach its target speed range – and the whole cycle will have to be repeated until all the components are working together, in harmony.

More changes may then be needed to the exhaust system and then all the carburettor settings will have to be finalized.

Each stage of this development is meant to go one step further towards the target performance but at each stage it is necessary to recognize where the engine is being restricted. The approximate effect of each modification is known and if it does not have the predicted effect, the reason must be found. For example, a more radical camshaft should raise peak power and cause a loss of low-speed power; if peak power is not increased and moved further up the speed scale, then something – possibly the intake system or the exhaust – is restricting the engine. It will not be possible to increase the power further until this restriction has been removed.

Eventually the limits of the engine will be reached. As far as power is concerned, the gains will be very small compared to the losses at lower speeds; the powerband will become too narrow to be usable. The engine will also become unreliable and eventually it will not be possible to fit stronger parts; there will be a finite limit.

If the development is done on a dynamometer it is fairly easy to see small changes which are not so clearly defined if the testing is done at a track. The pragmatic approach to this says that if you can't tell the difference, then there isn't any and when you no longer see improvements, you stop testing.

Unfortunately engines do not know about this, and when they are set up at the track there is a strong tendency to make them run slightly weak and slightly over-advanced. The reason is that in this condition they give the most crisp response and their power – in the short term – is as good as if the settings were perfect. If the engine is then used for a lengthy period, it tends to run too hot and may develop pre-ignition and burn a valve or a piston. Or it may overheat its plugs and lose power. Once the engine feels at its best, it is usually necessary to go one step richer and one step further retarded – unless it is being used for short blasts, such as drag racing or hill climbing, in which case there may not be time for any damage to happen.

While dyno testing is the quickest and most accurate way to get the right settings, the whole machine still needs to be track tested afterwards. There are many things which the dyno does not evaluate; the ground clearance of the exhaust system, the throttle response after heavy braking, the full range of part-throttle conditions, the useful width of the power band in relation to the available gear ratios and so on. It is not unusual for a race-tuned motor to take a season to develop fully, after the dyno testing has stopped.

It is important to realize exactly what is being tested and not to rely on subjective judgements. What feels best will usually be too weak or too far advanced; apart from that if the carburation is a total disaster at 7,000 rev/min but cleans up at 8,000, the way the motor bursts into life as it reaches the clean region will feel a lot more impressive than if it ran cleanly all the way through.

To set up a track test it is essential to control the approach to a particular speed/gear and to measure the speed reached by a particular point, or to time the machine between two points. There are a number of conditions which will make a bike go faster around a circuit, not all of them to do with the engine. For example, a change of wind direction, another bike just in front, another bike just behind, the rider learning how to go faster through a corner, and so on. If you are trying to find out about the engine, all of these other factors have to be shut out.

The hardest part is probably the carburation, because most items in the carburettor have some effect over all of the others. The safest way to test, from the engine's point of view, is to start too rich, and work down to the optimum setting, as rich mixtures will not cause anything to overheat. Once the carburettor is set approximately, get everything else set and leave the final carb settings until last.

Similarly, a cold plug grade should be used, changing to a hotter-running grade if the first plug fouls too easily. For this reason, do not take too much notice of plug readings – while a change to a lighter or darker colour indicates that the plug is running hotter or cooler respectively, it is not necessarily a sign of mixture strength or ignition advance.

Ignition timing should be retarded until there is a noticeable drop in throttle response and then **advanced a fraction further**. Carburation tests then fall into two categories, **WOT** at full revs on the fastest part of the circuit and whatever tests are **convenient** to run on the rest of the circuit. Keep reducing the main jet until performance reaches a maximum – making sure that the start of the straight is maintained at the same speed. Then go up one size.

If there is somewhere else that the engine is pulling WOT, from a lower speed, check its performance here – or set up an acceleration test timed between, say 6,000 and 8,000 rev/min. If a different main jet is needed to get the best performance here, the fuel slope is wrong (see Chapter 7). Use the best main jet for the low-speed test and, if this is smaller than the best in the high speed test, use a smaller air jet. If the low-speed test requires a bigger main jet, use it with a bigger air jet.

It is important to make only one change at a time and to concentrate on the main jet first – but this assumes that the engine is running tolerably well on the other systems. If not; if it is particularly obnoxious at half-throttle, say, then experiment with the needle settings, or whatever is necessary, to get it running acceptably. This will avoid possible engine damage and is necessary anyway, because a large change in the needle or needle jet will have some effect on the main jet requirements.

Having obtained an approximation on the main jet and air jet, re-set the idle adjustment, paying particular attention to any hesitation or flat spots just off idle, up to about 1/4 throttle. If there is a bad flat spot in this region, it may need a change of air slide, otherwise it can probably be cured by adjusting the pilot setting.

If the engine is showing any particular problems, make a note of the crank speed and the throttle position and devise some acceleration tests which use this setting – it will help to mark a scale on the twistgrip. The engine should, of course, be able to run at a constant speed anywhere within its power band, and be able to accelerate smoothly away from this speed. The throttle position will indicate which part of the carburettor is dominating the flow – but if large changes have to be made, it will also be necessary to go over all of the other settings again.

The part-throttle tests have to meet four basic conditions; small throttle opening/high engine speed; small throttle opening/low engine speed; large throttle opening/low engine speed; large throttle opening/high engine speed.

There is an infinite number of combinations of load and speed between these settings, but this will give the basic adjustment and any problems which occur later can be dealt with individually.

The combination of throttle and speed also has to be related to the engine's state of tune – a highly modified engine will not run at all well if the throttle is snapped open at low revs.

At low throttle, the top portion of the needle is the dominant factor but it is helped by the air slide cutaway, the pilot setting and the primary main jet, if one is fitted. If the needle setting works well at high speed but not at low speed, try using one of these additional factors to compensate.

At large throttle openings, the lower portion of the needle jet controls the flow, with some influence from the main jet. The pilot jet and the primary main continue to flow through the full range of throttle settings, so they will also add some fuel to the total.

Low-load running and throttle response can be altered on CV carburettors in several ways. (1) A heavier spring will hold the piston down longer, weakening the mixture because a higher region of the needle will control fuel flow. A lighter spring will let the piston lift earlier and will enrich the mixture. (2) A spacer under the piston will raise the fully closed position, reducing the air speed and reducing the fuel flow from pilot jet, by-passes and primary main jets, if fitted. (3) The vent hole(s) in the underside of the piston control the rate at which it lifts. They also provide a measure of damping, preventing the piston rising too quickly, rising too far and then falling back or fluttering under certain gas flow conditions. Making the hole larger will encourage the piston to lift more quickly and will reduce the damping effect. Making the hole smaller will slow down the piston's reaction to air speed changes and will give more damping or more stability. As with a stronger spring, holding the piston down tends to weaken the mixture at this point; lifting it further will enrich it.

If the mixture is good at steady speeds, but poor in transient conditions (engine pick-up, throttle response, acceleration) then the needle/needle jet are probably right; the response can be tuned by getting a different balance between spring strength and vent hole size. If piston flutter is a problem (unstable steady-speed fuelling, erratic response, surging) then it may be possible to get the same vent hole area but with two (or more) small holes instead of one.

When an air box or air filter has been removed or made less restrictive its effect on the pressure in the carburettor will not be the same – for the same engine speed and load, the pressure in the carburettor venturi will be greater (closer to atmospheric). The corrections necessary for jetting have already been described but, depending on where the CV carb's piston is vented, this increase in pressure will mean that the piston will not lift so far, weakening the mixture. In the same way, modifications to the air box or anything upstream of the carburettor can increase the intake pressure to the point where it will not be low enough to operate vacuum fuel taps. Usually this only happens at high load conditions, and then there is a delay as the fuel supply is used up, so the symptoms of fuel starvation/weak running only appear after several seconds' running. Where intakes have been seriously modified, manual on/off taps should be fitted as a matter of course.

CV carbs fitted at a steep downdraught angle (as on the OW01 and YZF Yamahas) can sometimes create problems when the inertia of the pistons makes them lift (i.e. travel forward) under heavy braking. When the throttle is opened for acceleration this causes a rich-mixture misfire, until the pistons have returned to their normal positions. Typically this only shows up at certain places – usually the type of hairpin bend which is approached by a fairly lengthy period of braking followed by progressive acceleration. With no damping mechanism on the air slides, it is difficult to find a solution which only works when the brakes are used. One last-resort remedy is to use the stop-light switch to cut out the fuel pump during braking.

Response to sudden throttle opening can be achieved by using an accelerator pump, while high speed weakness can be cured by fitting a high-speed jet (see Chapter 7).

It is quite likely that, having gone through the full range of adjustments, it will be possible to get a further improvement by repeating the entire process. At the end it is also possible that there will be one region where there is a pronounced flat spot (most four-cylinder engines suffer this in the 6,000 to 7,000 rev/min region, and the symptoms are made worse when the air box is removed and a 4–1 exhaust is fitted. It can be lessened by altering the carburation, but it cannot be entirely eliminated.

Gearing

Once the engine is giving maximum output it will be necessary to choose the best gearing to suit it. Unless a close ratio set of gears is available, it is unlikely that any of the intermediate gear ratios can be altered, but the overall gearing can be changed by using different gearbox and rear wheel sprockets.

For circuit use, top gear should be chosen to given maximum engine speed at the fastest part of the circuit; on many short circuits this will mean a very low final drive ratio and first and possibly second will never be used. An alternative is to raise the gearing so that the highest speed gives maximum revs in fourth (or fifth out of six) gear. As well as gearing for the highest speed, the number of gear changes around the circuit – and the places in which they have to be made – should also be considered. There is a calculation and a BASIC program in the Appendix which gives the engine thrust available in each gear and compares it to the air drag on a level road, to show the maximum speed available. It also demonstrates how this is altered by changing the gearing, the tyre size, and the power characteristics.

Appendix

Valve lift, velocity   143
Valve acceleration   144
Valve time-area   144
Road loads   150
Compression ration   161
Piston travel v. crank rotation   162
Piston velocity and acceleration   164
Vibration and balance   165
Cam timing v. cylinder head height   167
Valve guide clearance   167
Torque wrench settings   168
Strength of materials   168

This appendix contains data and calculations which relate to the above topics, plus programs, some of which are written in BASIC for microcomputers and others which are suitable for programmable calculators such as the Texas Instruments TI53.

The computer programs are written in BBC BASIC and BASIC 2.

The BASIC program A1 computes valve lift, valve velocity, valve acceleration and valve time-area, for a given engine specification. It also shows how these quantities vary when relevant engine dimensions are changed or when the engine speed is changed.

The data required by the computer is: bore, stroke, intake and exhaust valve diameter, number of valves per cylinder, valve seat angle, valve timing, valve lift at 10-degree intervals and engine speed range.

Valve lift

A conventional valve lift diagram is displayed, showing the phasing of the cams and the overlap period. The effect of changing the phasing can be seen instantly.

Valve velocity

For small angular changes, the instant valve velocity is v where

$$v = 6N(x_2 - x_1)/(c_2 - c_1) \times 10^{-3} \, \text{m/s}$$

where N = crank speed in rev/min
$x_{1,2}$ = valve lift in mm at $c_{1,2}$
$c_{1,2}$ = crank angle in degrees/relative to TDC
and $c_2 - c_1$ is small relative to the cam's duration

The peak velocity can be compared to the stock engine and to state of the art competition engines.

Valve acceleration
Using the above notation, the acceleration a, in m/s² is:

$$a = 6N(v_2 - v_1)/(c_2 - c_1) \times 10^{-3}$$

When the acceleration reaches a positive maximum, the contact stress between the cam and follower (and between the valve and valve lifter) will be greatest. This figure should be compared to previous figures to see how the stress will increase due to the change in acceleration. The actual stress will be $(F + ma)/A$, where F is the spring force at that particular lift, m is the mass of the moving parts, a is the acceleration and A is the contact area. So, an increase in 'a' can be countered by a reduction in m, or F, or an increase in A.

When the acceleration is negative, the cam is dropping away from the follower and valve motion is controlled by the spring. The required spring force to avoid valve float is Rx, where

$Rx > ma$

where R = spring rate
          x = valve lift
          m = mass of moving parts
          a = valve acceleration at lift x

(Note: m will be the weight divided by the gravitational constant, 32.2 ft/s² or 9.81 m/s², of the valve, collets, collar, and half of the spring(s). If a rocker arm is used then a factor for its moment of inertia should also be added or, to simplify things, its full weight added which will then allow a safety factor.) If the spring force Rx is less than ma then the valve will lose contact with the cam at that point. If Rx is significantly greater than ma, then the stress at the cam follower is unnecessarily high.

The program displays valve velocity and acceleration against a given crank speed. The result can also be varied by changing the valve lift data.

Valve time-area
Time-area is the quantity given by integrating the valve lift, diameter and the time it is open over each full cycle; it therefore represents the amount of air the port is able to flow. The program assumes that gas flow will not increase once the valve has exceeded a certain lift. If air flow tests, etc, indicate a particular value for this maximum effective lift, then it may be used in the program; otherwise the program will default to a value of 0.27 × valve diameter.

The program displays time-area in units of s-mm², and specific time-area, in units of s-mm²/cm³ (i.e. time-area per cc of piston displacement) so that the figures may be compared with engines of different sizes and so that the effect of changing the piston displacement may be seen.

The time-area figures are displayed against a range of crank speeds; alternatively, the figures can be displayed for a variety of valve diameters, at a constant crank speed.

Time-area values can be used in several ways, as shown in Chapter 5.

The effect of a different cam profile can be seen quite clearly in Fig. 70, which shows torque curves for a 608 cc Yamaha single, before and after fitting a factory sports cam. To match the new cam, bigger valves were fitted, along with modified carburettors, which gave more intake area and were jetted to run without the air cleaner. Table A.1 gives the complete change in specification.

Table A.1. Modifications to 608 cc Yamaha

Component	Stock	Modified
Intake valve dia., mm	36	37
Exhaust valve dia., mm	31	32
Intake tract, mm	29.6	32
Exhaust port, mm	31.5	31.5
Head volume, cc	68.5	70
Teikei carburettor:	(OE air cleaner)	(open bellmouths)
main choke, mm	24 × 29	24 × 31
CV choke, mm	26 × 30	29 × 33
main choke area, mm^2	572	620
CV choke area, mm^2	1077	1251
(slide) main jet/air jet	118/0.8	150/0.9
needle/needle jet	5C39/2.60	V10/2.60
pilot jet/air jet	46/0.6	50/0.7
(CV) main jet/air jet	100/1.3	155/0.8
needle/needle jet	5Z70/2.60	5Z70/2.60
Valve lift:		
intake, mm	8.7/110° ATDC	9.0/110° ATDC
exhaust, mm	8.5/100° BTDC	
Valve timing at 0.2 mm lift:		
intake opens	28° BTDC	60° BTDC
intake closes	68° ABDC	85° ABDC
exhaust opens	64° BBDC	96° BBDC
exhaust closes	32° ATDC	54° ATDC

Engine output v. intake time-area:

Crank speed rev/min	Torque, lb-ft/time-area, s-sq $mm/cc \times 10^{-3}$	
	Stock	Modified
5000	36.7/1.05	37.1/1.20
6000	33.7/0.87	38.7/1.00
6500	31.0/0.80	36.0/0.93
7000	28.5/0.75	33.0/0.86

Increase in intake area at carburettor	16%
Increase in valve time-area at 6000 rev/min	14.9%
Increase in torque at 6000 rev/min	14.8%
Increase in peak torque	5.4%
Increase in peak power	15%

Fig. 70. Torque curves for a 608 cc Yamaha, in standard form and with a sports camshaft, plus other modifications (see Table A.1). The points indicated show the relevant time-area values at those speeds (in s-sq mm/cc × 10^{-3})

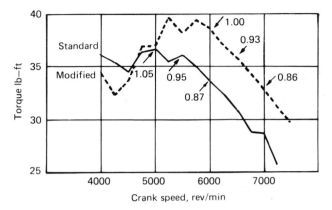

The increase in valve time-area had the typical effect of increasing high-speed torque and reducing low-speed torque. The results also show that the change in air flow/torque is closely related to the change in time-area, as calculated by program A.1. Some values for the stock engine and the modified version are shown in Table A.1 and in Fig. 70; peak torque occurs when the time-area is just over 1.00×10^{-3} s-sq mm/cc in both cases. And approximately the same level of torque occurs for equal time-area values except around peak torque. The increase in peak torque is possibly due to the air flow being boosted by resonant or ram effects in the intake or exhaust. This appears to happen at 5750 rev/min, at which speed the valves on the standard engine may not be open long enough for the engine to benefit from it or the effect could be reduced by the presence of the air filter.

Program A.1 Valve time-area

```
 10    REM TA4
 20    REM 28/10/85
 30    REM JWR
 40    DIM LI(36),LE(36),VI(36),VE(36),AI(36),AE(36),L(36)
 45    MODE 7
 50    CLS
 60    PRINT "Enter data in mm and"
 70    PRINT "crankshaft degrees."
 80    PRINT
 90    PROCINA
100    PROCINB
105    PROCINC
110    CLS
115    PRINT TAB(6,6)"For TIME-AREA.....press 1"
116    PRINT TAB(10,8)"VALVE LIFT..........2"
117    PRINT TAB(10,10)"VELOCITY............3"
118    PRINT TAB(10,12)"ACCELERATION........4"
120    PRINT TAB(7,14)"To VARY VALVE DATA......5"
122    PRINT TAB(10,16)"VARY CAM DATA.......6"
125    PRINT TAB(10,18)"STOP................7"
130    INPUT A
135    IF A=1 THEN PROCTA
```

```
 140   IF A=2 THEN MODE 4
 142   IF A=2 THEN PROCL
 145   IF A=2 THEN MODE 7
 150   IF A= 3 THEN PROCV
 155   IF A=4 THEN PROCA
 160   IF A=5 THEN PROCVALVE
 170   IF A=6 THEN PROCCAM
 175   IF A=7 THEN 199
 280   INPUT "if not known, enter 45. "Q
 285   Q=RAD(Q)
 290   C=PI*B^2*S/4000
 300   ENDPROC
 500   DEF PROCINB
 510   CLS
 520   PRINT "Adjust the valve clearances to their"
 525   PRINT "normal settings and turn the engine"
 530   PRINT "until the intake valve is lifted 1mm."
 540   PRINT "Measure the lift at 10 degree intervals"
 550   PRINT "until the valve closes."
 560   PRINT:PRINT "Enter the lift figures against the"
 570   PRINT "corresponding crank angle."
 580   PRINT "The input will stop when a lift "
 590   PRINT "value of 0 is entered."
 600   PRINT:PRINT
 610   PRINT "Crank angle BTDC"
 620   INPUT "for intake lifted 1mm? "IO
 630   PRINT
 640   PRINT TAB(5)"ANGLE"SPC(6)"LIFT"
 650   PRINT 0,1.0
 660   E=0:I=0
 670   REPEAT:E=E+10:I=I+1
 680     PRINT E:INPUT LI(I)
 690     VDU 11:VDU 11: PRINT E,LI(I)
 710   UNTIL LI(I)=0
 715   F=I
 720   ENDPROC
 725   DEF PROCINC
 726   CLS
 730   PRINT "Exhaust valve lift data:"
 735   PRINT:PRINT "Crank angle BBDC"
 740   INPUT "for exhaust lifted 1mm? "EO
 745   PRINT
 750   PRINT TAB(5)"ANGLE"SPC(6)"LIFT"
 755   PRINT 0,1
 760   E=0:I=0
 765   REPEAT:E=E+10:I=I+1
 770     PRINT E:INPUT LE(I)
 775     VDU 11:VDU 11:PRINT E,LE(I)
 780   UNTIL LE(I)=0
 785   G=I
 790   ENDPROC
1100   DEF PROCTA
1110   INPUT "Which valve, I/E? "A$
1120   PRINT:INPUT "Do you want to vary time-area against"
1130   PRINT TAB(6)"engine speed.....1"
1140   PRINT TAB(6)"valve size.......2"
1150   INPUT H
1170   IF H=1 THEN PROCSPEED
1180   IF H=2 THEN PROCSIZE
1190   ENDPROC
1200   DEF PROCSPEED
1210   CLS
1215   @%=&2020A
1220   INPUT "Minimum speed, rev/min? "N1
1230   INPUT "Maximum speed, rev/min? "N2
1240   INPUT "Increment? "N3
1250   PRINT:PRINT
1260   PRINT"Speed","Time-area","Specific"
1270   PRINT"rev/min","per cyl.","time-area"
1280   PRINT " ","s-sq mm","s-sq mm/cc"
1290   FOR I=N1 TO N2 STEP N3
1300       PROCCALC
1310       PRINT I,TA,TA/C*100" x10^-2"
1320     NEXT
1325   PRINT:PRINT:INPUT "To continue press C"E$
1326   IF E$="C" THEN 1330
1330   ENDPROC
```

```
2000  DEF PROCCALC
2010  IF A$="I" THEN 2070
2020  TA=0
2030  FOR X=0 TO G
2040    L(X)=LE(X):IF LE(X)>0.27*DE THEN L(X)=0.27*DE
2050    TA=TA+1.67/I*PI*L(X)*SIN(Q)*(DE-L(X)*COS(Q)*SIN(Q))
2060  NEXT
2062  TA=TA*NE
2065  GOTO 2120
2070  TA=0
2080  FOR X=0 TO F
2090    L(X)=LI(X):  IF LI(X)>0.27*DI THEN L(X)=0.27*DI
2100    TA=TA+1.67/I*PI*L(X)*SIN(P)*(DI-L(X)*COS(P)*SIN(P))
2110  NEXT
2115  TA=TA*NI
2120  ENDPROC
2200  DEF PROCL
2202  CLS
2205  LOCAL X,S,Z,I
2210  Z=0:REPEAT
2220    Z=Z+1
2230  UNTIL LI(Z)<LI(Z-1)
2240  LIM=LI(Z-1)
2250  S=800 DIV LIM
2260  FOR X=0 TO 1080 STEP 135
2270    MOVE X,0:DRAW X,800
2272    VDU 5
2275    IF X=270 THEN MOVE 240,840:PRINT "BDC"
2280    IF X=540 THEN MOVE 510,840:PRINT "TDC"
2282    IF X=810 THEN MOVE 780,840:PRINT "BDC"
2285  NEXT
2290  FOR X=0 TO 800 STEP 2*S
2300    MOVE 0,X:DRAW 1080,X
2305    IF X=2*S THEN PRINT "2.0mm"
2310  NEXT
2315  VDU 4
2320  I=INT((360-IO)*1.5)
2325  MOVE I,0
2330  FOR X=0 TO F
2340    DRAW 15*X+I,S*LI(X)
2350  NEXT
2360  I=INT((180-EO)*1.5)
2370  MOVE I,0
2380  FOR X=0 TO G
2390    DRAW 15*X+I,S*LE(X)
2400  NEXT
2410  Z=0:REPEAT
2420    Z=Z+1
2430  UNTIL LE(Z)<LE(Z-1)
2440  LEM=LE(Z-1)
2450  MOVE 900,0
2460  PRINT "TO ALTER PHASING PRESS P"
2470  MOVE 880,0
2480  INPUT "TO CONTINUE PRESS C"B$
2490  IF B$="C" THEN 2550
2500  CLS
2510  INPUT "INTAKE: ADVANCE, RETARD OR LEAVE A/R/L?"C$:IF C$="L" THEN 2530
2520  INPUT "HOW MANY DEGREES? "R:IF C$="A" THEN IO=IO+R ELSE IO=IO-R
2530  INPUT "EXHAUST: ADVANCE, RETARD OR LEAVE A/R/L?"D$:IF D$="L" THEN 2545
2540  INPUT "HOW MANY DEGREES? "RE:IF D$="A" THEN EO=EO+RE ELSE EO=EO-RE
2545  GOTO 2202
2550  A=0
2560  ENDPROC
2600  DEF PROCV
2610  LOCAL I
2620  FOR I=1 TO F
2630    VI(I)=(LI(I)-LI(I-1))/10
2640  NEXT
2650  FOR I=1 TO G
2660    VE(I)=(LE(I)-LE(I-1))/10
2670  NEXT
2680  CLS
2710  INPUT "Engine speed, rev/min?"N
2720  CLS:VDU 14
2730  PRINT "At "N" rev/min, valve velocities are:"
2735  @%=&2020A
```

```
2740    PRINT:PRINT "CRANK","INTAKE","EXHAUST"
2750    PRINT "ANGLE","(m/s)","(m/s)"
2760    FOR I=0 TO F
2770      PRINT I*10,VI(I)*6*N/1000,VE(I)*6*N/1000
2780    NEXT
2785    @%=10
2790    PRINT:PRINT:PRINT"PRESS ANY KEY TO CONTINUE"
2800    IF GET>1 THEN 2810
2810    ENDPROC
2900    DEF PROCA
2905    CLS
2910    LOCAL I
2915    INPUT "Crank speed, rev/min? "N
2920    FOR I=1 TO F
2930      VI(I)=6*N*(LI(I)-LI(I-1))/10
2940    NEXT
2950    FOR I=1 TO F
2960      AI(I)=6*N*(VI(I)-VI(I-1))*10^-7
2970    NEXT
2980    FOR I=1 TO G
2990      VE(I)=6*N*(LE(I)-LE(I-1))/10
3000    NEXT
3010    FOR I=1 TO G
3020      AE(I)=6*N*(VE(I)-VE(I-1))*10^-7
3030    NEXT
3035    PRINT:PRINT "At ";N" rev/min, the intake and "
3036    PRINT "exhaust valve accelerations are:":PRINT
3040    PRINT "Crank","Intake","Exhaust"
3050    PRINT "angle","(m/s^2)","(m/s^2)"
3055    @%=&2020A
3056    VDU 14
3060    PRINT
3070    FOR I=0 TO F
3080      PRINT I*10,AI(I),AE(I)
3090    NEXT
3100    @%=10
3110    PRINT:PRINT:PRINT "PRESS ANY KEY TO CONTINUE"
3120    IF GET>1 THEN 3130
3130    ENDPROC
3200    DEF PROCSIZE
3210    LOCAL S,T,Y
3220    S=DE:T=DI
3230    CLS
3240    INPUT "Crank speed, rev/min? "N
3245    I=N
3250    INPUT "Min valve diameter, mm? "D1
3260    INPUT"Max valve diameter, mm? "D2
3270    INPUT"Increment? "D3
3280    @%=&2020A
3285    PRINT:PRINT
3290    PRINT "Diameter","t-a","sp t-a"
3300    PRINT "(mm)","(s-sq mm)","(s-sq mm/cc)"
3310    FOR Y=D1 TO D2 STEP D3
3320      DE=Y:DI=Y
3330      PROCCALC
3340      PRINT Y,TA,TA/C*100" x10^-2"
3350    NEXT
3360    DE=S:DI=T
3370    @%=10
3380    PRINT:PRINT:PRINT "PRESS ANY KEY TO CONTINUE"
3390    IF GET>1 THEN 3395
3395    ENDPROC
4000    DEF PROCVALVE
4010    PROCINA
4020    ENDPROC
4100    DEF PROCCAM
4105    INPUT "Intake, exhaust or both? I/E/B "E$
4110    IF E$="E" THEN 4140
4120    PROCINB
4130    IF E$="I" THEN 4150
4140    PROCINC
4150    ENDPROC
```

Road loads

The thrust available at the back wheel depends on the engine torque, speed and the gear ratio between the crankshaft and the wheel. If the gearing lowers the speed by a factor of X, then it raises the torque by a factor of X; the gear ratios simply give us different values for X. The torque at the rear wheel multiplied by the wheel's rolling radius gives the force generated at the tyre's contact patch.

The force F (in lbf) is given by:

$$F = bhp \times 375/v$$

where
$v = KNG/W$
bhp = engine horsepower
v = road speed, mi/h
N = crank speed, rev/min
G = number of teeth on gearbox sprocket
W = number of teeth on wheel sprocket

(for shaft drive machines W/G = final reduction ratio)

and
$K = 0.00595R/PG_1$
P = primary reduction ratio
G_1 = internal gear ratio
R = tyre's rolling radius, inches

F can be plotted against v for the engine's full power curve, in each gear.

The BASIC program A.2 does just this, displaying the results graphically or in figures. The power figures can be obtained from dyno tests, manufacturers' quoted figures or magazine road tests. The effects of changing the power output or the gearing can then be seen quite clearly, particularly where the power band is made narrow or where there are large steps between the gears. The program also has a valve for air drag (assuming still air and level ground), so that the optimum gearing for maximum speed can be seen. The actual air drag will vary with the aerodynamics of the machine and its frontal area, so new factors can be applied at line 2290, based on measured performance.

Program A.2 Road loads

```
10 REM rl
20  REM  12/12/87:  17/4/89
25 REM Maximum speed 220 mi/h
30 REM BASIC2
40 REM jwr

100 GOSUB show: WINDOW #1 TITLE "RL.BAS"
101 REPEAT: rread=INKEY : UNTIL rread>-1
102 IF rread=114 THEN x=30: y=7: GOTO 140
103 CLS
105 INPUT AT(10;6)"Make/model: ",n$: GOSUB nam
106 PRINT: PRINT: PRINT AT(8) "This will be filed as "n1$". Is this OK? y/n";
```

```
107 REPEAT: nn=INKEY: UNTIL nn>-1: IF nn=110 THEN CLS: PRINT AT(10;2)"Full name: "n$:
    INPUT AT(10;6) "File name (max. 8 characters, no space): ",n1$: IF n1${-3 TO}<>"RL1" THEN
    n1$=n1$+".RL1"
110 PRINT: INPUT AT(10) "Number of gears: ",y
115 PRINT: PRINT AT(10) "For optimum speed, acceleration and gearshift"
116 PRINT AT (10) "predictions, give figures for the full engine speed"
117 PRINT AT (10) "range, up to maximum permissible speed.": PRINT
120 PRINT AT(10)"Number of power/speed entries:"
130 INPUT AT(10) "if less than 20, enter 0 ",x: IF x=0 THEN x=20

140 DIM n(x), hp(x), t(x), ld(x), v(y,x), vr(y,x), f(y,x), g(y): aa=0
145 DIM fav(y+1,220), warn(220), a(220), tim(220), thr(220),shift(y), vsh(y)
146 IF rread=114 THEN GOSUB rea: aa=1
148 REPEAT
150 GOSUB menu
160 UNTIL d>=10
200 END

300 LABEL menu
305 WINDOW #1 TITLE "RL.BAS"
310 IF aa=0 THEN GOSUB inpg: GOSUB inpp
315 aa=aa+1
320 CLS
330 PRINT AT(12;4)"Change gearing data.................1"
340 PRINT TAB(12) "Change power data...................2"
350 PRINT TAB(12) "Display power/torque................3"
360 PRINT TAB(12) "Display thrust/road speed...........4"
365 PRINT TAB(12) "Display acceleration *..............5"
368 PRINT TAB(12) "Display gearshift/traction data Ø...6"
370 PRINT TAB(12) "File data *.........................7"
380 PRINT TAB(12) "Read data from file.................8"
390 PRINT TAB(12) "Print data *........................9"
400 PRINT TAB(12) "STOP...............................10"
405 PRINT: PRINT: PRINT TAB(10)"* use item 4 before selecting this"
406 PRINT TAB(10)"Ø use items 4 and 5 before selecting this"
407 PRINT: PRINT TAB(10)"Use screen 2 to see current specification."
410 INPUT d: ON d GOSUB inpg, inpp, pow, thr, acc,warn, fil, rea, pri
430 RETURN

500 LABEL show
505 OPTION DATE 1
510 CLS
520 WINDOW #1 FULL ON
540 WINDOW #1 OPEN
541 ELLIPSE 4150;3000,2500,0.7 WIDTH 5 COLOUR 2
542 PRINT AT(30;6) POINTS(20);"Road Loads"
543 PRINT AT(8;10) ADJUST (16);"Acceleration and terminal speed predictor"
545 PRINT AT(12;18) "Do you want to enter new data"
546 PRINT TAB(12)"or read data from file? n/r "
550 RETURN

600 LABEL inpg
610 CLS: WINDOW #1 TITLE "DRIVELINE SPECIFICATION"
615 IF aa>0 THEN 720
616 d1$=DATE$
620 INPUT AT (10;2) "Primary reduction: ",p
630 FOR i=1 TO y
```

```
640 PRINT AT (10) "Internal gear ratio ";: INPUT ": ",g(i)
650 NEXT
660 INPUT AT (10) "Number of teeth on gearbox sprocket: ",t1
670 INPUT AT (10) "Number of teeth on wheel sprocket:   ",t2
680 PRINT "Do you know the rolling radius of the rear tyre? y/n "
690 REPEAT: e$=INKEY$: UNTIL e$>"":IF e$="n" THEN GOSUB tyre
700 PRINT:INPUT "Rolling radius of tyre (inches): ",r
710 IF aa=0 THEN 860
720 CLS: PRINT AT (10;4) "To change primary reduction press......1"
730 PRINT AT (21;5) "internal gears..............2"
740 PRINT AT (21;6) "gearbox sprocket............3"
750 PRINT AT (21;7) "wheel sprocket..............4"
760 PRINT AT (21;8) "tyre radius.................5"
775 PRINT AT (10;10)"No change..............................6"
780 INPUT bb: ON bb GOTO 790, 800, 810, 820, 830, 870
790 PRINT AT (10;12) "Current primary reduction is "p;: INPUT "New ratio: ",p: GOTO 860
800 FOR i=1 TO y: PRINT AT (10) "Current gear "i" is "g(i);: INPUT "New ratio: ",g(i): NEXT: GOTO 860
810 PRINT AT (10) "Current gearbox sprocket is "t1;: INPUT "New sprocket: ",t1: GOTO 860
820 PRINT AT (10) "Current wheel sprocket is "t2;: INPUT "New sprocket: ",t2: GOTO 860
830 PRINT AT (10) "Current rolling radius is "r: INPUT AT(10) "New radius: (enter 0 to see tyre sizes) ",r: IF r=0 THEN GOSUB tyre: IF r=0 THEN 830 ELSE 860
860 PRINT: PRINT AT (8) "Any other changes? y/n ", :REPEAT: ee$=INKEY$: UNTIL ee$>"": IF ee$="y" THEN 720
870 GOSUB spec: RETURN


900 LABEL inpp
910 CLS: WINDOW #1 TITLE "ENGINE OUTPUT"
920 PRINT AT (4;2) "Enter speed (rev/min) and engine output (bhp)."
980 PRINT "Enter speed; RETURN; output; RETURN. Enter 1 to stop input."
990 PRINT: PRINT AT (15) "Rev/min" AT (25) "bhp" AT (35)" torque, lb-ft"
1000 w=0
1010 REPEAT
1020 PRINT TAB(15) ">";:INPUT "", n(w);: IF n(w)=1 THEN 1030
1025 PRINT TAB(25)">";:INPUT"",hp(w);: t(w)=hp(w)*5252/(n(w)): PRINT TAB(35) ROUND(t(w),1)
1030 w=w+1
1040 UNTIL n(w-1)=1 OR w=x OR hp(w-1)=1
1042 PRINT: PRINT "Are the figures OK? y/n"
1043 REPEAT: e$=INKEY$: UNTIL e$>"": IF LOWER$(e$)="n" THEN CLS: GOTO 980
1050 RETURN


1500 LABEL calc
1510 FOR j=1 TO y
1520   q=p*g(j)*t2/t1
1530 FOR i=0 TO w-2
1540   v(j,i)=0.00595*r*n(i)/q:  vr(j,i)=ROUND(v(j,i))
1550 f(j,i)=hp(i)*375/v(j,i):       REM vr() is rounded to whole mi/h
1560 NEXT:                          REM for use in calculating av force
1570 NEXT
1580 RETURN


1600 LABEL thr
1610 GOSUB calc
1620 CLS: WINDOW #1 TITLE "THRUST (lbf) v ROAD SPEED (mi/h)"
1630 PRINT AT(10;4) "Figures or graph? f/g ",: REPEAT: th$=INKEY$: UNTIL th$>"": IF aa<3 THEN 1640: IF th$="f" THEN 1640
```

```
1632 PRINT TAB(10) "Drag factor: option 1,2,3 or 4"
1634 INPUT AT(10) "or select 5 to see drag data: ",ad
1640 IF th$="g" THEN GOSUB gra: GOTO 1740
1650 CLS
1660 PRINT TAB(10) "Road speed" TAB(25)"Thrust"
1665 PRINT TAB(16) "mi/h" TAB(28) "lbf"
1670 PRINT
1680 FOR j=1 TO y
1690 FOR i=0 TO w-2
1700 PRINT USING "             ##.#";,v(j,i),f(j,i): GOSUB wait
1710 NEXT
1720 PRINT: PRINT TAB(8) USING "##.##&"; v(j,i-1)*1000/n(i-1) " mi/h per 1000
rev/min": PRINT: PRINT
1730 NEXT
1735 PRINT TAB(20)"Press any key to continue."
1736 IF INKEY$="" THEN 1736
1740 RETURN


1700 LABEL wait
1710 IF ww>0 THEN ww=ww+1: GOTO 1770
1720 yy=YPOS
1730 IF yy>750 THEN 1780
1740 PRINT AT(10;20) "Press SPACE bar to continue.";
1750 IF INKEY$<>" " THEN 1750
1760 ww=1: PRINT AT (10;20)"                      "
1770 IF ww=13 THEN ww=0: PRINT: GOTO 1740
1780 RETURN


1800 LABEL gra
1810 GOSUB drag
1815 WINDOW #1 TITLE "Thrust (lbf) v. road speed (mi/h)"
1820  sx=7000/v(y,w-2):  sx=sx/1.05
1830 i=0
1840 j=-1
1850 j=j+1: IF j=w-2 THEN 1880
1860 IF f(1,i)>=f(1,j) THEN 1850
1870 IF f(1,i)<f(1,j) THEN i=i+1: GOTO 1840
1880 sy=4000/f(1,i): i2=i : sy=sy/1.05
1890 k=20: i=0: kk=ROUND(f(1,i)/50): kk=kk*10
1900  REPEAT: LINE (1000+k*i*sx);1000,(1000+k*i*sx);5000: i=i+1
1910  UNTIL k*(i-1)*sx>=5500
1920 FOR j=0 TO i-1: MOVE (800+k*j*sx);700: PRINT j*k: NEXT
1930 i=0
1940  REPEAT: LINE 1000;(1000+kk*i*sy),8000;(1000+kk*i*sy): i=i+1
1950  UNTIL kk*sy*(i)>=4000
1960 FOR j=0 TO i-1: MOVE 400;(875+kk*j*sy): PRINT j*kk: NEXT
1970 FOR i=1 TO y
1980 FOR j=0 TO w-3
1990  LINE       1000+v(i,j)*sx;1000+f(i,j)*sy,1000+v(i,j+1)*sx;1000+f(i,j+1)*sy
2000 NEXT
2010 NEXT
2020 i=0: REPEAT: dr=a+b*i+c*i^2
2030 PLOT 1000+i*sx;1000+dr*sy MARKER 5 SIZE 1 COLOUR 1
2040 i=i+1: UNTIL dr*sy>3800 OR i>=v(y,w-2)
2045 MOVE 1200;1500: PRINT "Final drive: "t2"/"t1
2046 MOVE 1200;400: PRINT n$": "d1$
2050 MOVE 5500;400: PRINT "Press c to continue."
```

153

```
2060 IF INKEY$<>"c" THEN 2060
2070 RETURN


2100 LABEL drag
2102 IF ad>0 AND ad<4 THEN 2270
2105 WINDOW #1 TITLE "Drag factors."
2110 CLS
2120 PRINT AT(6;2) "Overall drag is assumed to take the form"
2130 PRINT AT(6) "    dr = a + bv + cv2"
2140 PRINT AT(6) " where a,b and c are constants accounting for rolling and"
2150 PRINT AT(6) "driveline drag, aerodynamic drag, inertia of the"
2160 PRINT AT(6) "machine and inertia of the rotating parts.""
2170 PRINT
2180 PRINT AT(6) "For large, unfaired bikes (Z1000J, GSXII00)"
2190 PRINT AT(6) "try values of a=16, b=0 and c=0.0105       = option 1"
2195 PRINT
2200 PRINT AT(6) "For large, faired bikes (GSX-R1100)"
2210 PRINT AT(6) "try values of a=16, b=0 and c=0.0091       = option 2"
2215 PRINT
2220 PRINT AT(6) "For small, faired bikes (TZR250)"
2230 PRINT AT(6) "try values of a=16, b=0 and c=0.008        = option 3"
2235 PRINT
2240 PRINT AT(6) "Or new values can be used.                 = option 4"
2250 PRINT
2260 INPUT AT(10) "Option: ",ad
2270 IF ad=1 THEN a=16 : b=0: c=0.0105
2270 IF ad=2 THEN a=16 : b=0: c=0.0091
2280 IF ad=3 THEN a=16 : b=0: c=0.008
2290 IF ad=4 THEN INPUT AT(10)"Value for a: ",a;: INPUT AT(30)"Value for b: ",b;: INPUT AT(50)"Value for c: ",c
2300 CLS: GOSUB spec
2310 RETURN

2400 LABEL acc
2405 IF f(1,1)=0 THEN GOSUB calc: GOSUB drag
2410 GOSUB inpw
2420 GOSUB avg
2430 CLS: t=0: i=0: WINDOW #1 TITLE "Speed (mi/h) v Time (s)"
2440 REPEAT
2480   a(i)=0.682*thr(i)/m
2490   tim(i)=1/a(i): t=t+tim(i): REM t() in seconds
2495 IF i=60 THEN t60=t
2500 i=i+1:UNTIL  i-1=v(y,w-2)  OR  thr(i)<=0:vmax=i-1: etmax=t: j=i-1
2510 sx=6800/t:  sy=4000/(vmax+10):  i=0
2515 IF etmax>38 THEN n=4 ELSE n=2
2516 IF etmax>76 THEN n=8
2520 REPEAT
2530   LINE  1000+2*i*sx;1000,1000+2*i*sx;5000
2540   MOVE  800+n*i*sx;700: PRINT n*i
2550   i=i+1
2560 UNTIL  n*i*sx>=7000
2570 i=0
2580 REPEAT
2590   LINE  1000;1000+i*sy,8000;1000+i*sy
2600   MOVE  450;900+i*sy: PRINT i
2610   i=i+20
2620 UNTIL i*sy>=4000
```

```
2625 dist=0: et=0: et4=0: d4=0: sp=0
2627 gr=sg
2630 FOR i=0 TO vmax-1
2640   dist=dist+(tim(i)*(2*i+1)/2)/3600:  et=et+tim(i)
2650   LINE  1000+et*sx;1000+i*sy,1000+(et+tim(i+1))*sx;  1000+(i+1)*sy
2655 IF shift(gr)=i THEN 2656 ELSE 2660
2656   LINE  1000+et*sx;1000+i*sy,1050+et*sx;820+i*sy
2657   gr=gr+1
2660 IF dist>0.23 AND et4=0 THEN GOSUB quart: PRINT d4: PLOT 1000+et4*sx;1000+v4*sy MARKER 2
2670 NEXT
2680 dmax=dist: IF dist>0.2494 THEN 2685
2681 REPEAT: dist=dist+(tim(i)*i)/3600: et=et+tim(i):UNTIL  dist>0.2499
2682 et4=et: v4=i-1: REM for bikes which reach max speed inside 1/4mile
2684 dmax=ROUND(dmax,2):vmax=ROUND(vmax,1): etmax=ROUND(etmax,1):
et4=ROUND(et4,2):v4=ROUND(v4,1):t60=ROUND(t60,2)
2690 MOVE 4500;2000: PRINT "Max speed: "vmax"mi/h in "etmax"s "
2700 MOVE 6500;1750: PRINT "and "dmax" mi"
2710 MOVE 4500;1500: PRINT "SS 1/4  mi: "et4"s/"v4"mi/h"
2720 MOVE 4500;1250: PRINT "0-60mi/h:    "t60"s"
2725 MOVE 900;400: PRINT n$": "d1$
2730 MOVE 5800;400: PRINT "Press c to continue."
2740 IF INKEY$<>"c" THEN 2740
2750 RETURN

2900 LABEL inpw
2910 CLS: WINDOW #1 TITLE "DIMENSIONS AND LAUNCH DATA"
2912 IF wt=0 THEN 2920
2914 PRINT AT (10) "Is the data (weight, wheelbase, etc) the same as before? y/n: ",
2914  REPEAT: inp$=INKEY$: UNTIL inp$>"": IF inp$="y" THEN 3007
2920 INPUT AT(10) "Weight of bike and rider, lbf: ",wt:m=wt/32.2: PRINT
2930 PRINT TAB(10) "If the following data is not available,"
2940 PRINT TAB(10) "enter 0; the program will assume a typical value."
2950 PRINT
2960 INPUT AT(10) "Wheelbase, inches: ",wb: IF wb=0 THEN wb=56
2970 PRINT TAB(10) "Centre of gravity co-ordinates"
2980 INPUT AT(10) "Horizontal distance from rear wheel spindle, inches: ",cgx: IF cgx=0 THEN cgx=wb*0.5
2990 INPUT AT(10) "Height above ground, inches: ",cgy: IF cgy=0 THEN cgy=2*r+3: PRINT
3000 INPUT AT(10) "Coefficient of friction, tyre/road: ",mu: IF mu=0 THEN mu=1: PRINT
3002 PRINT AT(10)"(Note: to start in a gear other than first, alter line 3215.)":  INPUT
AT(10) "At what engine speed is the clutch engaged? ",cl: IF cl=0 THEN cl=n(0)
3003 IF cl>0 AND cl<n(0) THEN CLS: PRINT AT(10) "This is less than the first speed value.":
PRINT AT(10) "Enter 0 or a value greater than "n(0): GOTO 3002
3005 PRINT AT(10) "At what engine speed are gearshifts made? "
3006 INPUT AT(10) "If 0 is entered, the program will calculate optimum speeds. ",gs
3007 CLS: GOSUB spec
3010 RETURN

3100 LABEL avg
3105 FOR i=0 TO 220: warn(i)=0: thr(i)=0: FOR ii=1 TO y: fav(ii,i)=0: NEXT: NEXT
3110 i=1
3120 REPEAT: j=0
3130 REPEAT
3135    FOR v=vr(i,j) TO vr(i,j+1)
3140        fav(i,v)=f(i,j)+(f(i,j+1)-f(i,j))*(v-vr(i,j))/(vr(i,j+1)-vr(i,j))
3150    NEXT
```

```
3160    j=j+1
3170  UNTIL j-1=w-2
3200  i=i+1:IF i-1=y THEN 3210
3205  IF vr(i-1,j-1)<vr(i,0)-1 THEN GOSUB slip
3210  UNTIL i-1=y:
3215  sg=1: REM gear in which start is made
3218     speed=CEILING(cl*0.00595*r/(p*g(sg)*t2/t1))
3220  FOR v=0 TO speed
3225    fav(sg,v)=fav(sg,speed):  warn(v)=warn(v)+0.5
3226    IF fav(sg,v)>0.8*f(sg,i2) THEN fav(sg,v)=0.8*f(sg,i2):
3227  NEXT
3228  IF gs<>0 THEN GOSUB shift: GOTO 3340: REM condition for pre-chosen gearshift rpm.
3230  i=sg: v=0
3240   REPEAT
3250    REPEAT
3252      thr(v)=fav(i,v)
3255   GOSUB wheel
3260       thr(v)=thr(v)-(a+b*v+c*v^2)
3270     v=v+1
3275     UNTIL fav(i,v-1)<=fav(i+1,v-1) OR v-1=vr(i,w-2)
3280       shift(i)=v-1:  vsh(i)=shift(i)*p*g(i)*t2/(t1*0.00595*r)
3290     i=i+1
3295   UNTIL i-1=y OR thr(v-1)<=0
3340 RETURN

3400 LABEL shift
3410  i=sg:v=0
3420  FOR gear=i TO y-1: shift(gear)=ROUND(gs*0.00595*r/(p*g(gear)*t2/t1)):
vsh(gear)=gs: NEXT
3430 REPEAT
3440 REPEAT
3441 IF fav(i,v)=0 THEN fav(i,v)=f(i,0): IF FRAC( warn(v))<>0.5 THEN warn(v)=warn(v)+0.5
3442    thr(v)=fav(i,v)
3445 GOSUB wheel
3450    thr(v)=thr(v)-(a+b*v+c*v^2):   v=v+1
3460  UNTIL v-1=shift(i)
3470  i=i+1
3480 UNTIL i-1=y OR thr(v-1)<=0
3490 RETURN

3500 LABEL slip
3510  FOR v=vr(i-1,j-1) TO vr(i,0)-1
3520    fav(i,v)=f(i,0):  warn(v)=warn(v)+0.5
3530 NEXT
3540 RETURN

3600 LABEL tyre
3610 WINDOW #2 FULL ON: CLS #2: WINDOW #2 TITLE "Tyre size and rolling radius": WINDOW #2 OPEN
3620 SET ZONE 19
3625 PRINT #2 " Rolling radius in inches ±2%, based on ETRTO standard."
3630 PRINT #2: PRINT #2 "  15-inch","17-inch","18-inch","18-inch"
3640 PRINT #2
3650 PRINT #2 "140/90    11.96","2.50     10.73","3.60     11.62","100/80    11.66"
3660 PRINT #2 "150/90    12.30","2.75     11.11","4.10     12.11","110/80    11.96"
3670 PRINT #2 " ","3.00     11.38","4.25     12.91","120/80    12.26"
3680 PRINT #2 "  16-inch","4.50     12.59","4.25/85  12.46","130/80    12.57"
```

```
3690 PRINT #2 " "," "," " "

3700 PRINT #2 "4.60      11.42","100/80    11.18","90/90      11.7","110/70    11.55"
3710 PRINT #2 "100/90    11.07","120/80    11.79","100/90    12.04","140/70    12.34"
3720 PRINT #2 "110/90    11.42"," ","110/90    12.38","150/70    12.61"
3730 PRINT #2 "120/90    11.75","110/90    11.90","120/90    12.72",""
3740 PRINT #2 "130/90    12.09","120/90    12.24","130/90    13.06"," "
3750 PRINT #2 "140/90    12.44","130/90    12.59","140/90    13.40","170/60    12.49"
3760 PRINT #2 "100/80    10.70"," "
3770 PRINT #2 "120/80    11.30","140/80    12.39"
3780 PRINT #2 "150/80    12.21"
3790 PRINT #2 TAB(40) "Press c to continue."
3800 IF INKEY$<>"c" THEN 3800
3810 WINDOW #1 OPEN
3820 RETURN

4000 LABEL pri
4050 GOSUB setup
4010 LPRINT n$, d1$: LPRINT
4020 LPRINT "Final drive "t2"/"t1: LPRINT
4030 LPRINT "speed, mi/h","thrust, lbf": LPRINT
4040 FOR j=1 TO y
4050 FOR i=0 TO w-2
4060 LPRINT  ROUND(v(j,i),1),ROUND(f(j,i),1)
4070 NEXT: LPRINT "Gear "j": "  ROUND(v(j,i-1)*1000/n(i-1),1)" mi/h per 1000rev/min":
LPRINT
4080 NEXT
4090 LPRINT: LPRINT
4100 RETURN
4850 LABEL pow
4860  pmax=hp(0):tmax=t(0):i=0
4870 REPEAT: i=i+1: pmax=MAX(pmax,hp(i)): UNTIL i=w-1: i=0
4875 REPEAT: i=i+1: tmax=MAX(tmax,t(i)): UNTIL i=w-1: mmax=MAX(pmax,tmax)
4880  sx=6500/n(w-2)
4890  sy=3800/mmax
4900 CLS: WINDOW #1 TITLE "Output (bhp and torque, lb ft) v. crank speed, rev/min xl0^3"
4910 k=1000:  i=0
4920 REPEAT:  LINE  1000+k*i*sx;1000,1000+k*i*sx;5000:  MOVE  800+k*i*sx;600:  PRINT
i:  i=i+1
4930 UNTIL  k*(i-1)*sx  >=6500
4940 i=0: IF sy<390 THEN k=10 ELSE k=1
4950 REPEAT:  LINE  1000;1000+k*i*sy,8000;1000+k*i*sy:  MOVE  400;1000+k*i*sy:  PRINT
i*k:  i=i+1:  UNTIL  k*i*sy>=3900
4960 FOR i=0 TO w-3: LINE
1000+n(i)*sx;1000+hp(i)*sy,1000+n(i+1)*sx;1000+hp(i+1)*sy:  NEXT
4970  FOR  i=0  TO  w-3:  LINE  1000+n(i)*sx;1000+t(i)*sy,1000+n(i+1)*sx;1000+t(i+1)*sy:
NEXT
4980 MOVE 1000;300: PRINT n$": "d1$
4990 MOVE 5800;300: PRINT "Press c to continue."
5000 IF INKEY$ <>"c" THEN 5000
5010 RETURN

6000 LABEL quart
6010 et4=et: v4=i: d4=dist: counter=0: ii=i
6020 REPEAT
6030 v4=v4 + 0.1: counter=counter + 1
6040 et4=et4 + 0.1/a(ii)
```

```
6050   d4=d4+ v4/(a(ii)*36000)
6060 IF counter=9 THEN counter=-1 AND ii=ii+1
6070 UNTIL d4>0.2498
6080 RETURN

6100 LABEL nam
6110 n1$="": n2$=".RL1"
6120 le=LEN(n$)
6130 FOR i=1 TO le
6140   t$=n${i}
6150 IF ASC(t$)=32 THEN t$=""
6160 n1$=n1$+t$
6170 NEXT
6180 le=LEN(n1$)
6190 IF le>8 THEN n1$=n1${TO 3}+n1${-5 TO}
6195   n1$=n1$+n2$
6200 RETURN

6300 LABEL fil
6310 CLS: WINDOW #1 TITLE "Saving data on disc file"
6320 PRINT AT(10;3)"Put a formatted disc in drive B:"
6330 PRINT: PRINT AT(10)"The data will be stored in \RL\"n1$
6340 PRINT: PRINT AT(10)"To alter the filename, press A or press"
6342 PRINT AT(10) "any other key to continue."
6350 REPEAT:f1$=INKEY$: UNTIL f1$>"": IF LOWER$(f1$)="a" THEN GOSUB ns
6360 DRIVE "B"
6370 ON ERROR GOTO 6550
6380 CD \RL\
6400 ON ERROR GOTO 0
6420 IF n1$="" THEN n1$=n3$
6430 IF FIND$(n1$)>"" THEN REPEAT: GOSUB rname: UNTIL FIND$(n1$)="" OR r$="y"
6450 OPEN #3 OUTPUT n1$
6460 PRINT #3,n$
6462 PRINT #3,d1$
6470   PRINT  #3,w,y,p,t1,t2,r,wt,wb,cgx,cgy,mu,cl,gs
6480 FOR i=1 TO y
6490   PRINT #3,g(i)
6500 NEXT
6510 FOR i=0 TO w-2
6520   PRINT #3,n(i),hp(i),t(i)
6530 NEXT
6540 CLOSE #3: GOTO 6570
6550 IF ERR=133 THEN GOSUB drctry
6560 RESUME NEXT
6570 RETURN

6600 LABEL rname
6610 PRINT
6620 PRINT AT(10)"A file "n1$" already exists."
6630 INPUT AT(10)"Do you want this file to replace it? y/n ",r$
6640 IF r$="y" THEN 6670
6650   mm=ASC(n1${-1})+1
6660 n1${-1}=CHR$(mm)
6670 RETURN

6700 LABEL drctry
6710   MD \rl\
```

```
6720  CD \rl\
6730 RETURN


6800 LABEL rea
6810 CLS: WINDOW #1 TITLE "READ DATA FROM FILE"
6820 DRIVE "B"
6830 PRINT AT(10;6)"Put the file disc into drive B:"
6840 PRINT AT(10)"Press a key when ready."
6850 IF INKEY$="" THEN 6850
6855 ON ERROR GOTO 6990
6860 CD \rl\
6870 ON ERROR GOTO 0
6880 PRINT: PRINT AT(10)"Note that file names are compressed into a maximum": PRINT AT(10)"of eight characters, with no spaces, and have ": PRINT AT(10)"an extension .RLx":GOSUB look
6890 PRINT: REPEAT: INPUT "Filename for machine, or Q to quit: ",n3$
6895 IF n3$="q" THEN 6905
6900 IF FIND$(n3$)="" THEN PRINT "No file for "n3$
6905 UNTIL FIND$(n3$)>"" OR n3$="q": IF n3$="q" THEN 7000
6910 PRINT: PRINT "Reading "n3$
6920 OPEN #3 INPUT n3$
6930 INPUT #3,n$
6932 INPUT #3,d1$
6940   INPUT  #3,w,y,p,t1,t2,r,wt,wb,cgx,cgy,mu,cl,gs
6950 FOR i=1 TO y
6960 INPUT #3,g(i)
6965 NEXT
6970 FOR i=0 TO w-2
6975  INPUT #3,n(i),hp(i),t(i)
6980 NEXT
6985 CLOSE #3
6986  m=wt/32.2
6987 CLS:PRINT: PRINT AT(10)n$
6988 tim=TIME:REPEAT:UNTIL TIME>tim+500:GOTO 7020
6990 IF ERR=133 THEN PRINT AT(10)"There is no \RL\ directory on this disc."
7000 PRINT TAB(10)"Try another disc or quit? a/q ";: INPUT re$
7010 IF re$="a" THEN 6810
7020 RETURN


7100 LABEL ns
7110 PRINT "Existing filename: "n1$
7120 INPUT "New filename, including extension: ",n1$
7130 PRINT "Existing full name: "n$
7140 INPUT "New name: ",n$
7150 RETURN


7200 LABEL wheel
7210 IF  thr(v)*cgy>wt*cgx  THEN  warn(v)=  warn(v)+1:  thr(v)=wt*cgx/cgy
7220  IF  thr(v)>wt*mu  THEN  warn(v)=warn(v)+2:  thr(v)=wt*mu
7230 RETURN

7400 LABEL warn
7405 WINDOW #1 TITLE "Gearshift and traction data"
7410 CLS
7415  sxx=6300/vmax
7420 FOR i=0 TO vmax
7430 IF INT(warn(i))=1 OR INT(warn(i))=3 THEN
```

```
       LINE(1500+i*sxx);1000,(1500+(i+1)*sxx);1000 WIDTH 5 COLOUR 1
7440   IF INT(warn(i))=2 OR INT(warn(i))=3 THEN
       LINE(1500+i*sxx);1500,(1500+(i+1)*sxx);1500 WIDTH 5 COLOUR 2
7450   IF FRAC(warn(i))=0.5 THEN LINE(1500+i*sxx);2000,(1500+(i+1)*sxx);2000 WIDTH 5
       COLOUR 11
7460   NEXT
7470   LOCATE 1;17: PRINT "Wheelie"
7480   LOCATE 1;15: PRINT "Wheelspin"
7490   LOCATE 1;13: PRINT "Clutch slip"
7500   FOR i=0 TO vmax STEP 10
7510      LINE(1500+i*sxx);1000,(1500+i*sxx);2100
7520   MOVE (1300+i*sxx);700: IF i<vmax-15 THEN   PRINT i
7530   NEXT
7540   LOCATE 10;1: PRINT "Gearshift    ","rev/min","mi/h"
7545   IF gs=0 THEN PRINT AT(10) "(calculated)"
7550   PRINT
7560   FOR i=1 TO y-1
7570   PRINT AT(10) i" to "i+1,ROUND(vsh(i)),shift(i)
7580   NEXT
7580   PRINT AT(12;19)"Road speed, mi/h" AT(50;19) "Press any key to continue"
7590   IF INKEY$="" THEN 7590
 7600  RETURN
8000   LABEL look
8010   PRINT "Here are the files on this disc:"
8020   FILES
8030   RETURN

8100   LABEL setup
8110   LPRINT CHR$(15)
8120   LPRINT CHR$(27)+"l"+CHR$(15)
8130   LPRINT CHR$(27)+"N"+CHR$(5)
8140   RETURN

8200   LABEL spec
8202   WINDOW #2 SIZE 27,30: WINDOW #2 PLACE 400,0
8203   WINDOW #2 TITLE "CURRENT SPECIFICATION"
8210   CLS #2: PRINT #2
8220   PRINT #2 " Final drive       "t2"/"t1
8230   PRINT #2 " Primary           "ROUND(p,3)
8240   PRINT #2 " Roll radius       "r
8250   PRINT #2
8260   PRINT #2 " Wheelbase         "wb
8270   PRINT #2 " cgx - r wheel     "cgx
8280   PRINT #2 " cgy               "cgy
8290   PRINT #2 " Total weight      "wt
8300   PRINT #2 " μ                 "mu
8310   PRINT #2
8320   PRINT #2 " Start gear        "sg
8330   PRINT #2 "Clutch dump rpm "cl
8340   PRINT #2
8350   PRINT #2 " Total drag"
8360   PRINT #2 "    =a + bv + cv2"
8370   PRINT #2 "        a="a
8380   PRINT #2 "        b="b
8390   PRINT #2 "        c="c
8400   RETURN
```

The same calculation can be made on a programmable calculator, such as the Texas Instruments TI53, as shown in program A.3.

Program A.3 Road loads (TI53)
First calculate K and G/W to four decimal places.

Step	Key	Symbol		Step	Key	Symbol	
00	55	×	input N rev/min	12	85	=	displays v in mi/h
01	83			13	41	STO	
02	()	K1	enter K to four	14	81	R/S	input bhp
			decimal places	15	55	×	
03	()	K2		16	74	3	
04	()	K3		17	52	7	
05	()	K4		18	63	5	
06	55	×		19	45	÷	
07	83			20	51	RCL	
08	()	G/W	enter G/W to four	21	85	=	displays F in lbf
			decimal places	22	81	R/S	
09	()	G/W		23	31	2nd	
10	()	G/W		24	81	RST	
11	()	G/W					

Compression ratio

Program A.4 uses the TI53 to calculate the change in compression ratio caused by machining an amount h from the gasket face of the cylinder head or barrel (or by adding a thickness h in the form of an extra gasket, etc, in which case h should be negative in the program).

The calculation for the compression ratio C is

$$C = (V_c + V_{sw})/V_c$$

and the new ratio C_1 is:

$$C_1 = (V_c \pm Ah + V_{sw})/(V_c \pm Ah)$$

where V_c = combustion chamber volume
V_{sw} = volume displaced by piston
h = depth of material removed from gasket face
A = area of cylinder bore

Program A.4 Compression ratio (TI53)
Calculate V_c and A to four digits and V_{sw} to five digits.

Step	Key	Symbol		Step	Key	Symbol	
00	55	×	input h	06	()	V	enter V_c
01	()	A	enter A	07	()	V	
02	()	A		08	()	V	
03	()	A		09	()	V	
04	()	A		10	85	=	
05	65	−		11	41	STO	

Step	Key	Symbol	
12	65	–	
13	()	VSW	enter V_{sw}
14	()	VSW	
15	()	VSW	
16	()	VSW	
17	()	VSW	
18	85	=	

Step	Key	Symbol	
19	45	÷	
20	51	RCL	
21	85	=	displays
			compression ratio
22	81	R/S	
23	31	2nd	
24	81	RST	

Input any value of h and the program will display the new compression ratio. If h is a thickness being added to the gasket face then enter –h.

Piston travel v. crank rotation

It is often useful to be able to convert crankshaft degrees into piston height before TDC. This calculation does it and the following TI53 program will display piston height for any angle which is fed into the calculator. From this a graph of piston height v crank angle can be plotted, allowing instant conversion from degrees to mm and vice versa.

The TI53 is restricted to a program length of 31 steps and consequently the figures used in it have to be rounded off to a restricted number of digits – including one for the decimal point.

$$d = \frac{S}{2} + L - \frac{S}{2}\cos X - L \sin\left[\cos^{-1}\left(\frac{S}{2L}\sin X\right)\right]$$

where: d = distance before TDC
S = stroke
L = length of connecting rod
X = angle before TDC in degrees

This expression can also be written:

$$d = \frac{S}{2}\left\{1 + \frac{2L}{S} - \cos X - \sqrt{\left[\left(\frac{2L}{S}\right)^2 - \sin^2 X\right]}\right\}$$

Program A.5 Piston travel (TI53)

1. Calculate S/2L to three digits including point (=A)
2. Calculate S/2 to four digits including point (=B)
3. Calculate S/2 + L to five digits including point (=C)

Note that d will be negative.

Step	Key	Symbol	
00	41	STO	input X (deg B/ATDC)
01	22	sin	
02	55	×	
03	83	.	enter A
04	()	A	
05	()	A	
06	85	=	
07	31	2nd	
08	23	$\cos^{-1}$	
09	22	sin	
10	55	×	
11	()	L	enter L
12	()	L	
13	()	L	
14	75	+	
15	()	B	enter B
16	()	B	
17	()	B	
18	()	B	
19	55	×	
20	51	RCL	
21	23	cos	
22	65	−	
23	()	C	enter C
24	()	C	
25	()	C	
26	()	C	
27	()	C	
28	85	=	displays d (mm B/ATDC)
29	81	R/S	
30	31	2nd	
31	81	RST	

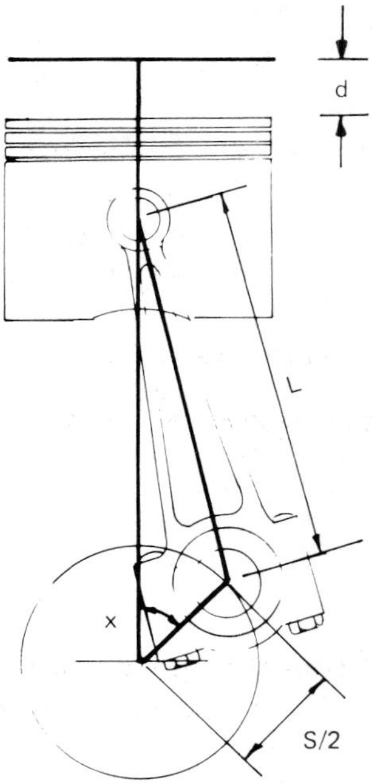

Fig. 71. Dimensions used in the program which relates piston height to crank rotation

Program A.6 Piston travel

```
10 REM PT
20  REM  24/8/85
30 REM BBC B
40 REM PISTON HEIGHT v CRANK ROTATION
45 REM JWR
50 CLS
60 PRINT: PRINT "Enter data in mm and"
65 PRINT "crankshaft degrees, B/ATDC"
70 INPUT "Stroke: "S
80 PRINT "Rod length: "
90 INPUT "If not known, enter 0: "L: IF L=0 THEN L=2*S
100 INPUT "Maximum crank angle: "X1
110 INPUT "Minimum crank angle: "X2
120 INPUT "Incremental step: "X3
130 CLS
135 VDU 14: REM prevents screen scrolling until SHIFT is pressed
140 @%=&2020B: REM print format: 2 decimal places, column width 11 (hex B) characters
145 ON ERROR GOTO 370
150  A=S/2+L
160 Z1=RAD(X1): Z2=RAD(X2): Z3=RAD(X3)
170 PRINT SPC(6) "Angle",SPC(3) "Distance"
180 PRINT SPC(3) "from TDC", SPC(2) "below TDC"
190 FOR Z=Z1 TO Z2 STEP Z3
200    P=S/2*COS(Z)
210    R=S/2*SIN(Z)
220    Q=SQR(L^2-R^2)
230    D=A-P-Q
249   PRINT DEG(Z), D
250  NEXT
260 PRINT TAB(23) "Press any key"
265 PRINT TAB(25) "to continue"
270 IF GET>0 THEN 280
280 CLS
285 PRINT:PRINT:PRINT
290 PRINT "To alter the STROKE press............1"
300 PRINT: PRINT SPC(13) "ROD LENGTH.....2"
310 PRINT: PRINT SPC(13) "CRANK ANGLE...3"
320 PRINT: PRINT SPC(10) "To STOP ..........4"
330 INPUT M: CLS
340 IF M=1 THEN PRINT "Stroke:" S "mm": INPUT "New stroke:" S: GOTO 150
350 IF M=2 THEN PRINT "Length:" L "mm": INPUT "New length:" L: GOTO 150
360 IF M=3 THEN 100
365 IF M>3 THEN 380
370 IF ERR=21 THEN PRINT "The rod must be longer": PRINT "than the crank throw."
380 END
```

Piston velocity and acceleration

The piston's mean velocity is v, where

$$v = 2sN$$

where s = stroke
       N = crank speed

The generally accepted maximum for plain compression rings is 20 m/s.

The inertia force generated by the piston is the product of its mass and its acceleration and it can be important to know how this alters when the piston mass or the engine speed is changed. If the acceleration is a at a point where the crank makes an angle X to the line of the stroke, then

$$a = b^2 r \left(\cos X + \frac{r}{L} \cos 2X \right)$$

where: b = crank velocity in radians per second
 r = stroke/2
 L = connecting rod length

the acceleration will be a maximum at TDC, when X = 0

$$a_{max} = b^2 r \left(1 + \frac{r}{L} \right)$$

(Note that b = 0.1047N, where N is the engine speed in rev/min). A change in the reciprocating mass will, if the rod and big end are not strengthened, cause the safe maximum crank speed to be reduced. If the original redline was N_1 then the new one will be N_2 where

$$N_2 = N_1 \sqrt{M_1/M_2}$$

where M_1 and M_2 represent the original and the new mass of the piston, pin, rings, circlip, small end and the top two-thirds of the connecting rod.

Vibration and balance

A single piston generates inertia forces in the reciprocating parts which are proportional to the acceleration (a above), and has a constant rotating force caused by the mass of the big end, crank web and the lower portion of the connecting rod. The rotating force can be balanced by incorporating an equivalent mass on the opposite side of the crank throw.

The reciprocating force has two factors, proportional to cos X and cos 2X; as X increases from zero to 360 degrees, cos X will vary from +1 to −1 at 180 degrees, and back to +1 at 360 degrees, i.e. it reverses at the same frequency as the engine's rotational speed. The other factor, cos 2X, becomes −1 at X = 90 and +1 at X = 180, i.e. it reverses at twice the engine frequency. These two terms are sometimes known as the primary and secondary forces. At mid-stroke (X = 90) the primary force is zero, but the secondary force is at a maximum. At BDC the primary force is at a negative maximum (i.e. acting downwards) while the secondary force is at a positive maximum (i.e. acting upwards) so the total inertia force is proportional to the primary minus the secondary. However, at TDC they are both at a positive maximum – both acting upwards – and both will be added together to give the total, so the inertia force at TDC is greater than that at BDC (which is why a 180-degree twin does not have as good balance as its layout might suggest).

The secondary force is much smaller than the primary; depending on the proportions of the stroke and the rod length it will be in the order of 25 per cent of the primary.

A weight can be added to the flywheel opposite the crankpin so that it produces a centrifugal force B, which will oppose the primary out-of-balance (OOB) force. The ratio of B to the primary force is called the balance factor; if this is 100 per cent then the primary force will be balanced at TDC and BDC, leaving the secondary force unopposed. But, at midstroke, the force B will be acting horizontally with nothing to balance *it*. The vibration will be as severe as an unbalanced engine, but in a horizontal instead of a vertical sense.

Where counterweights are used to balance a single piston, the balance factor is usually between 50 and 90 per cent, the optimum amount being determined experimentally to suit the particular installation.

One or two balance shafts can, as Fig. 72 shows, balance the primary forces (if the balancer is driven at engine speed; to balance the secondary forces it would have to be driven at twice engine speed).

On motorcycle engines the secondary force is generally neglected, except where it is automatically balanced by the layout of the engine, e.g. horizontally-opposed twins and fours.

The force generated by counterweights is proportional to the mass and the radius of gyration; to increase the counterweight it is usually necessary to drill or machine metal from the opposite side of the web or shaft and the

Fig. 72. Functions of a single balance shaft and (bottom) two balance shafts. The out-of-balance force created by the piston is shown as P, and that of the big-end assembly as B

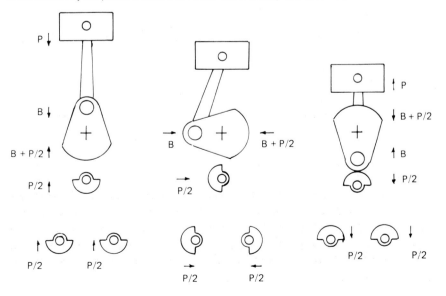

greater the distance of this from the axis of the crank or the balance shaft, the greater the effect it will have.

For example, if the weight of the piston is increased by a weight x, then the extra primary OOB force produced by this will be xr (r = half the stroke). It can be countered by machining a weight y from the crankpin side of the flywheel, at a radius p, so that yp = xr.

Where an engine has one balance shaft, an amount equal to xr/2 should be machined from the crankshaft and a similar amount from the unweighted side of the balancer.

Where an engine has two balance shafts, then an amount of xr/2 must be added to each balance weight or subtracted from the unweighted side.

Cam timing v head height

If overhead camshafts are driven by chain or by belt and the cylinder head is lowered to raise the compression ratio then the cam timing will be retarded because the driving run of the chain-belt will have been shortened.

The change in timing, measured in degrees of crankshaft rotation, will be

$$D = \frac{720h}{NP}$$

where D = crank angle (degrees)
        h = change in cylinder head height
        P = pitch of chain
        N = number of teeth on cam sprocket

A TI53 program will display the crank angle for any input of 'h' (if the head is raised, the cam will be advanced).

Program A.7 Cam timing (TI53)

Step	Key	Symbol		Step	Key	Symbol	
00.	55	×	input h	08	83	P	enter P
01	52	7		09	()	P	
02	73	2		10	()	P	
03	82	0		11	()	P	
04	45	÷		12	()	P	
05	()	N	enter N	13	85	=	displays D
06	()	N		14	81	R/S	
07	45	÷		15	31	2nd	
				16	81	RST	

Valve-guide clearance

Several manufacturers now check the valve guide wear in service by fitting a new valve into the guide, pushing it down to a repeatable position, e.g. where the collet groove aligns with the top of the guide and measuring the rock of the valve by placing a dial gauge against its head, at right angles to the stem. The rock should be checked in two positions, at right angles to one another.

Assuming some uniformity in the wear, this can be related to the actual valve-to-guide clearance by:

$$x = R.L_g/(2L_v + L_g)$$

where x = clearance between valve and guide
R = total rock
L_g = length of guide
L_v = length from end of guide to measuring point

Program A.8 Valve-guide clearance

This program displays the clearance for any input of R:
First calculate $L_v/L_g = z$

Step	Key	Symbol		Step	Key	Symbol	
00	45	÷	input R	08	( )	Z	
01	43	(		09	75	+	
02	73	2		10	72	1	
03	55	×		11	44	)	
04	( )	Z	enter Z	12	85	=	displays x
05	( )	Z		13	81	R/S	
06	( )	Z		14	31	2nd	
07	( )	Z		15	81	RST	

Torque wrench settings

In the absence of specified torque settings, the following chart provides a guide for the materials used in most Japanese roadsters. The threads should be clean and dry; ordinary bolts have no marking or sometimes have a figure 4 on their head; high-tensile bolts may have a figure 7.

bolt diameter (mm)	ordinary N-m (lb-ft)	high tensile N-m (lb-ft)
4	1–2 (0.8–1.5)	1.5–3.0 (1–2.2)
5	2–4 (1.5–3.0)	3.0–6.0 (2.2–4.4)
6	4.0–7.0 (3.0–5.2)	8.0–12.0 (6.0–8.8)
8	10–16 (7.4–11.8)	18–28 (13.3–20.7)
10	22–35 (16–26)	40–60 (30–44)
12	35–55 (26–40)	70–100 (52–74)
14	50–80 (37–59)	110–160 (81–118)
16	80–130 (59–96)	170–250 (125–184)
18	130–190 (96–140)	200–280 (148–206)
20	190–250 (140–184)	250–300 (184–220)

To convert N-m into kg-m multiply by 0.1020.

Strength of materials

The properties of some materials make them more suitable than others for a given job, but there is more to it than simply being strong or weak because materials often have different strengths in different directions – and they can be loaded in tension, compression, torsion, shear or bending (or some combination of several of these). In addition, metals and particularly steels,

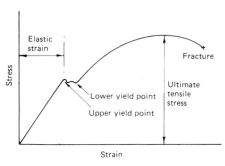

Fig. 73. Stress/strain relationship for steel

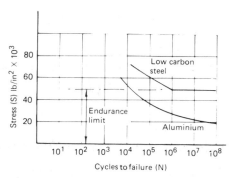

Fig. 74. Typical fatigue test results on steel and aluminium

are – up to a point – elastic; that is they return to their original dimension when the load is removed. Beyond this point (called the yield point) they become plastic; that is, the deformation is permanent. But heat treatment can restore them to their original state, so they can be drawn into shape, beaten or forged into shape and still retain their original, elastic property and strength. Some forming processes, such as forging, work-harden the surface, leaving a tough outer layer. Other heat treatment processes such as nitriding, cause salts to be dissolved into the outer layer of metal, again forming a tough skin.

Stress is the force applied to the object, divided by its original cross-section area. *Strain* is the ratio of the change in dimension compared to the original dimension before the stress was applied. The ratio of stress divided by strain is a measure of the material's strength, called its modulus of elasticity, or Young's modulus.

If a sample of the material is held in a rig so that an increasing stress can be applied to it and the resulting strain can be measured, then it will go through a phase of elastic strain, during which the strain is proportional to the stress. It then reaches its *elastic limit*, at which point plastic deformation begins; brittle materials do not show much elastic strain and no plastic strain, on reaching their elastic limit they break with a sudden *brittle fracture*, the uniformly jagged edges of which show that all of the cross-section has failed at the same time.

Steels show a different characteristic. At the elastic limit they *yield*, that is the material suddenly stretches, giving a large increase in strain for no increase (and perhaps a decrease) in stress. The metal structure reorganises itself and further stress produces a progressively increasing amount of strain – which is plastic deformation. The material work-hardens or strain-hardens, being able to withstand a stress well above its original yield point. Finally the stress reaches a maximum at the material's *ultimate tensile strength*, at which point no further increase in stress is necessary to produce an increase in strain. The sample may form a 'neck' across its weakest point, and in this place the cross section will be reduced, so further strain will result without any extra load being added; the local stress will have increased and will stretch the metal, but the total, measured, stress will actually decrease up to the point where the material fails. The *ductility* of the material is the amount of plastic strain it can withstand before fracturing.

In a tensile test such as this the load is applied steadily and similar tests can be made to evaluate the material's performance in compression, shear, etc. The stress to strain ratio represents its strength while the area under the graph represents the energy absorbed and is a measure of the material's *toughness*.

When parts are used on engines they are loaded in similar ways and their material and size is selected from the results of this type of test, to cope with the forces which the part will have to contain. However, if the load fluctuates or reverses (as the force generated by a rotating crankshaft would) then another condition appears, called fatigue. Parts which are easily capable of withstanding the force under steady conditions, will weaken after a few million cycles (or less) and then fail.

A different type of test is needed to evaluate this property; usually the sample is rotated while a steady load is applied to one side of it and a counter records the number of revolutions. A switch controlling the drive is attached to the sample so that when it fails it switches off the motor. The stress and the number of cycles to failure can then be plotted as a graph – called an S-N graph.

As the stress is reduced the number of cycles to failure (N) is increased and the graph usually forms a curve, becoming asymptotal at a very low level of stress for materials like aluminium alloys (see Fig. 74). The implication is that, if the component is only moderately loaded then it will, eventually, suffer fatigue failure.

However, steels show a different characteristic, which is emphasized if the N scale is plotted as an exponential, as in Fig. 74. The smooth curve suddenly flattens in the region of 10^6 to 10^8 cycles. The level of stress at which this occurs is called the material's endurance limit and tests suggest that, as long as the stress does not exceed this limit, then the material will never fatigue. This is quite a useful feature in an engine which can log up more than half a million cycles in one hour.

For all materials, the fatigue strength is improved (i.e. the S-N curve is shifted up the S axis) in the following conditions:

1. If the surface is polished to remove small cracks.
2. If the surface is hardened by chemical or mechanical means, which produce compressive stresses in the surface.
3. If a steady, compressive stress is superimposed on the cyclic stress.

Fig. 75. Stress raiser. (*Left*) a notch or a hole (a crack is regarded as an ellipse in which b is very much greater than c). (*Right*) a change in section. Maximum stress occurs where the section begins to expand from diameter D_2. A radiussed fillet alleviates the problem and the greater the radius R, the less stress-raising effect there will be. In any case R should be greater than $0.2D_2$

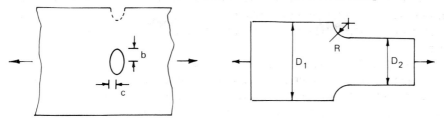

Fig. 76. Stress-relief. (a) A radius or a stress-relieving groove in the shoulder of a shaft. (b) Increasing the section of a shaft where it is carried in the bearing. (c) Threads and splines, etc, are also stress raisers – the shaft or the shank of the bolt/stud should be slightly less than the root diameter of the thread, etc. (d) Highly-loaded female threads in castings should have the first one or two threads removed and the entrance to the hole radiussed. (e) A groove (*top*) with the largest possible radius can be used to stress-relieve a bolt or splined shaft where the root diameter is less than the shaft diameter. Where a groove has to be made in a highly-loaded part like a valve stem (*bottom*) it should also be fully radiussed. (f) Threaded fasteners can be stress-relieved with a radius at the entry (*top*) or specially-constructed parts (*bottom*) can be used

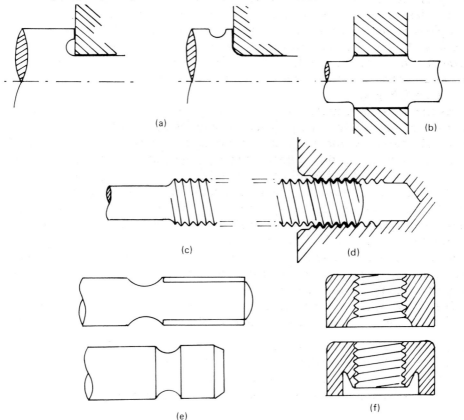

The fatigue strength is reduced if:

1. The surface has a plated coating:
2. A steady, tensile load is superimposed on the cyclic stress.
3. The material is in a corrosive environment.
4. The surface or the shape of the part has a *stress raiser*.

A stress raiser is a sudden change of section or shape which causes a local increase in stress. As most of the stress in a loaded part is carried near the surface, a crack, notch, or hole will not be able to support any stress, which will then be diverted to the neighbouring material. In this one place the stress concentration may be high enough to exceed the endurance limit. So while the part is perfectly strong enough to carry the load, it will fatigue locally after a few million cycles and the crack will extend, reducing the available cross-section of material still further. For the same load, this raises the level of stress and the failure begins to accelerate.

Typically a fatigue failure will show 'scallop' marks, where a crack has propagated from a stress-raiser, but the part has continued to function until the reduction in area has brought the stress up to its yield point. This will be followed by a brittle fracture or a ductile fracture (in which there is considerable plastic deformation), depending on the material and the manner in which it is loaded.

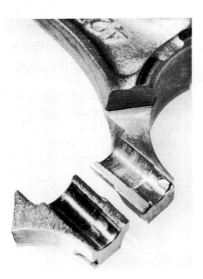

Fig. 77. A connecting rod sectioned through the big-end bolt hole, showing the square edge left when the rod was machined to accept the head of the bolt. The corner should have been radiussed

Fig. 78. A rod from the same engine as Fig. 77, which had already failed on one side. Sectioning the other side and using crack-detecting dye revealed the beginnings of a crack, starting at the stress-raising, square notch (arrowed)

The stress concentration at a hole is approximately three times the stress level applied to the component (if the hole is small compared to the size of the component). If the hole is not circular but elliptical, the increased stress S_{max} is given by: $S_{max} = S(1 + 2b/c)$ where b and c are the radii of the ellipse. This is leading up to the effect of a crack, whose end can be considered as the tip of a very long and thin ellipse in which b is very much greater than c...

Note that the long axis of the ellipse (b) has to be at right angles to the stress, which is why cracks at right angles to the force direction spread rapidly. If b is much smaller than c, then $S_{max} \rightarrow S$.

A change in section obviously increases the stress where the section is reduced but maximum stress occurs at the point at which the section begins to increase. The ratio of this to the applied stress is called the stress concentration factor. If the change in section is made by radiussed fillets, experiments have shown that the stress concentration factor increases as the fillet radius gets smaller. When the radius is less than $0.2\times$ the width of the smaller side of the part, the stress concentration rises dramatically. It reaches a factor of $3\times$ the applied stress when the fillet is about $0.05\times$ the material width.

Some examples of stress-raisers and methods of stress-relieving parts are shown in Figs. 75 and 76.

Shot peening improves the fatigue strength of materials because the surface hammering causes internal compressive stresses in the surface layer and there is also hardening caused by plastic deformation. The intensity of the peening depends upon the shot material, which may be iron, steel or glass, and whose effect diminishes as the angle of impact moves from 90°.

Components with smooth surfaces will have their fatigue strength improved by 10 to 20%, while those with rough or 'defective' surfaces may have their fatigue strength improved by as much as 50 to 70%. The intensity of shot peening is measured on the Almen scale in which the letters, N, A and C refer to low-, medium- and high-intensity, respectively, while the figures refer to the first two digits of the measure of the increase in curvature (in inches $\times 10^{-3}$). (e.g. 14N = $\times$ 0.35 mm or 0.0138 inch)

The shot-peen treatment recommended for steel components by the Italian Standards Institute is:

Thickness of part, mm	Intensity (Almen grade)
3–5	14N
5–10	12A
10–30	18A
>30	11C

The shot size has to take into account the shape of the component, oilways, etc. and the permitted abrasive effect.

Shot diameter, mm	Surface finish, Ra, m
0.2	0.8
0.4	1.6
0.8	3.0
1.5	12.0

Finally, temperature has a marked effect on the strength of metals. To show a comparison between different metals (which melt at different temperatures) an *homologous* temperature is used. This is the ratio of the absolute temperature of the material to the absolute temperature of its melting point. At an homologous temperature of 0.2 (i.e. 20 per cent of its melt temperature) steel weakens quite suddenly; at an homologous temperature of 0.3 its Young's modulus is only 2/3 of its original value. Aluminium shows a similar weakening at an homologous temperature of about 0.55; at 0.7 its Young's modulus is reduced by 50 per cent and for pure aluminium this would occur in the region of 370 deg C.

Index

A/R ratio, 9
Absorption silencer, 80
Acceleration,
 piston, 165
 valve, 143
Accelerator pump, 96
Advance curve, 106
Advanced ignition, 106
Air bleed, 85
 box, 97
 cooling, 118
 filter, 97, 98
 flow, 99
 rig, 20
 test, 52, 58
 intake, 97
 jet, 85, 91
 shifter, 122
 slide, 88, 93
Air/fuel ratio, 82, 108
Almen scale, 173
Alternator, 126
Altitude control, 97
Amal jets, 92
 needle jet, 95
 needle, 95
 smoothbore, 101
Amplifier, 111
Appendix, 143–174
A/R ratio, 9
Atomizer, 94
Avgas, 13, 107

Baffles, crankcase, 117
Balance factor, 165
 shaft, 126, 166
 carburettor, 92
 crankshaft, 125
Ballast resistor, 111
Base circle, 68
Bearings, 123, 125
 load capacity, 127
Bellmouth, 98
Bernoulli's theorem, 84
Big bang engines, 34
Big-bore engines, 34
Big-end bolts, 128
 clearance, 127
Blow-by, 125
Blower - see Supercharger
Bluegas, 13
Blueprinting, 121
Boost, 8
Bore gauge, 133
 measurement, 131
Bore/stroke ratio, 57
Boring bar, 19
Bosch dynamometer, 24
Brake - see Dynamometer
Breaking-in - see Running-in
Breather, 117, 125
 cam box, 136
Brittle fracture, 169
Brown Boveri, 11
Bucket and shim, 69, 136
Built-up cranks, 128
Burette, 18
Burn, 84

Cagiva piston, 131
Cam, 35, 51–72, 66
 box breather, 136
 chain tensioner, 72
 drive, 72, 136, 167
 follower, 65, 69
 phasing, 66
 profile, 145
 reprofiling, 68
 timing, 167
Capacitive discharge, 105
 silencer, 80
Carburettor air flow, 99
 balance, 92
 downdraught, 94
 float, 91
 main system, 91
 needle, 95
 jet, 94
 part-throttle, 140
 size, 57
 smoothbore, 99
 theory, 84
 tuning, 90, 139
Carillo, 129
Cast pistons, 38

Catch tank, 97
CDI, 105
Ceramics, 49
CFR engine, 13
Choke, 98
Circlips, piston pin, 132
Close ratio gears, 121
Clutch, 122
 spring, 123
Coil, 111
Cold-start jet, 98
Colortune plug, 92
Combustion chamber, 44
 temperature, 107
Compression, 5
 ratio, 46, 161
 rings, 43
 tester, 18
Comprex, 11
Connecting rod, 128
 failure, 172
 length, 129
Constant vacuum - see CV
Constant velocity - see CV
Contact-breaker, 105
Contact stress, 65, 71, 143
Cooling, 114–120
 system, bleeding, 120
Cosworth, 28
 piston, 131
Counterweights, 166
Crankcase, 36, 124
 breather, 97, 117, 125
 distortion, 126
Crankshaft, 125
 modification, 126
 runout, 127
 speed, 165
Cross-connected ports, 59, 99
CV carburettor, 88, 93, 140
Cylinder, 38–50
 block, 132
 head, 45, 135
 height, effect on cam timing, 167
 measurement, 131

pressure, 108
studs, 133

De-aeration chamber, 125
Degree disc, 17
Degreeing-in, 66
Dell'Orto, 93
 accelerator pump, 96
 needle, 95
 jet, 94
Detonation, 5, 103, 106, 107
Development, 27–37
Dial gauge, 17, 128
Diffuser, 73, 79
Digital ignition, 106
Dowels, 125
Downdraught carburettor, 94, 141
Dry clutch, 122
 sump, 116, 118
Ducati 600/750, 29
 851/888 air intake, 97
Ductility, 170
Duration, 55
Dwell, 105
Dynamometer, 20
Dynojet dynamometer, 25

Eddy current dynamometer, 24
Elastic limit, 169
Electrical tester, 18
Emulsion tube, 87, 94
Endurance limit, 169
Engine breather, 117
 building, 121
 configurations, 27–33
 friction, 14
 testing, 20
EPROM, 102
Exhaust collector, 78
 cross pipes, 79
 pipe length, 77
 port, 75
 process, 6
 system, 73–81
 4-into-1, 74, 142

175

4-into-2-into-1, 74
Expansion stroke, 6
Fatigue, 40, 124, 134, 170
  failure, 172
  test, 169, 170
Feeler gauge, 18
Final drive, 142
Fine tuning, 137–142
Firing interval, 33
Five-valve head, 48, 53, 57
Flat-4, 32
Flat slide, 88
  spot, 140
  twin, 29
Float chamber, 95
  height, 91
Floating bearing, 11
Flow meter, 83
Flywheel, 166
Forged pistons, 10, 38
Four-valve head, 45, 57
Friction losses, 14, 49, 121
Froude G4 dynamometer, 24
Fuel, 11
  flow, 91
  gallery, 101
  head, 96
  injection, 101
  level gauge, 18
  map, 101
  pump, 94, 96
  slope, 86
  stand-off, 98
  starvation, 96, 141
  system, 82-104
  tap, 91

Gas-filled ring, 133
Gasket, 134
Gear ratios, 142, 150
Gearbox, 121
Generator - see Alternator
Go/no-go tool, 18
Gudgeon pin - see Piston pin
Guides, 61

Hall effect, 106
Head - see also Cylinder head
Head gasket, 46, 134
  volume, 46
Header pipe, 75, 78
Heat, 1
  treatment, 169
Heenan & Froude, 20, 24
Helmholtz chamber, 80
Hemi-head, 44
Hepolite Apex ring, 43
High speed jet, 96
Hitachi turbo-charger, 12
Homologous temperature, 174
Honda NR500/750, 32
  R&D, 14
  RCB dry sump, 116
  RSC dry clutch, 122
  V-4, 31, 130
Honing, 41, 132
Horsepower, 2
HT lead, 105, 111

Idle system, 88, 92
Ignition, 105–113
  amplifier, 111
  coil, 111
  distributor, 122
  timing, 106, 108, 112
  trigger, 106, 112
In-line three, 29
  four, 30
  five, 32
  six, 33
Inductive discharge, 105
Inertia dynamometer, 25
  forces, 126
  piston, 165
  valve train, 71
Injector nozzle, 101
Intake length, 58
  silencer, 97
  stack (trumpet, bellmouth), 98
  surge tank, 99
Internal micrometer, 133

Jet kit, 91
  air, 85

main, 85
  needle, 85
Kawasaki, 23
  close ratio gears, 122
  GPX750, 14
  GPz550 power fade, 117
  Z250, 72
  ZXR750, 57
  air box, 97
Keihin semi-flat slide carburettor, 104
Knock - see also Detonation
Knock sensor, 10

Laurence-Scott dynamometer, 14, 25
Line filter, 91, 96
Liner - see also Cylinder
Liners, 38–50
Liquid cooling, 119
Losses, 14, 121
Lubrication, 114–118
  splash, 129
Lucas RITA, 112

Magnetic trigger, 106
Main jet, 85
  baffle, 96
  extension, 96
  types, 91
Mass flow, 3
Megaphone, 73
Methanol, 11
Micrometer, 17, 131
Mikuni jets, 92
  needle, 95
  jet, 94
  smoothbore, 102
Mixture loop, 83
Modulus of elasticity, 169
MON, 13
Motoring test, 14
MTBE, 13

N_2O - see Nitrous oxide
Napier ring, 43
Needle, 88, 95
  jet, 85, 94
  valve, 95

NGK, 108
Nikasil, 41
Nitriding, 169
Nitro-methane, 11
Nitrous oxide, 12, 103
Norton P86, 28
Notch effect, 172

Octane booster, 12
  rating, 13, 107
Ohm-meter, 18
Oil breather, 116
  consumption, 116
  control, 115
  cooler, 114, 117
  cooling, 7, 49, 63, 114
  flow, big end, 129
  pressure tester, 18
  pump, 116, 124
  scraper, 117
  ring, 43
  trap, 125
  viscosity, 118
OOB - see Out of balance
Optical trigger, 106
Out of balance force, 166
Oxidising agent, 103
Oxygenates, 13

Parallel twin, 27
Part-throttle carburation, 140
Pegged bearing, 125
Pent-roof head, 45
Pilot jet, 95
Piston, 38-50, 130
  acceleration, 164
  area, 39
  clearance, 41
  CV carburettor, 88
  inertia, 165
  mass, 42
  pin, 132
  proportions, 131
  ring, 43, 47
    clearance 132
  rock, 49
  skirt clearance, 47
  speed, 42
  travel, 162
  program, 164

-to-head
  clearance, 44
valve
  (carburettor),
  93
  velocity, 164
Plastic strain, 169
Plastigage, 18, 127
Plug - see spark
  plug Ports, 56,
  59
Power, 2
  fade, 117
  jet, 96
  loss, 14
Pre-ignition, 107,
  135
Pressed cranks, 128
Pressure regulator,
  101
  relief valve, 115
  wave, 4, 59, 73
Pressurized intake,
  97
Primary choke, 94
  main jet, 89, 93
  vibration, 165
PTFE pad, 132
Pulse, 4, 59, 73
Pulser coil, 106
Pumping losses, 14,
  121

Rajay, 8
Rear wheel thrust,
  150
Reciprocating mass,
  36, 130, 165
Regrinding cams,
  68
Resonance, 73, 146
Restrictor jet (oil),
  114
Retarded ignition,
  106
Rich mixture, 82
Ring lands, 132
Rings, 43
Rise time, 105
Road loads, 150
  program, 150
Rocker ratio, 69
Rod - see
  Connecting rod
Roller bearings,
  123, 128
Rolling road
  dynamometer,
  25
RON, 13
Rotax single

cylinder, 77
Running-in, 16,
  121

S-N graph, 170
Scavenge pump,
  118
Schenck
  dynamometer,
  23
Secondary main jet,
  89
  pipe, 79
  vibration, 165
Selective assembly,
  16
SFC, 24, 82, 86
Shell bearing, 127
Shot peening, 19,
  125, 129, 173
Silencer, 79–80
  intake, 97
Single cylinder, 27
Sintered metal, 49
Slide carburettor,
  88
Slipper piston, 49
Small end, 132
Smoothbore
  carburettor, 99
Spark plug, 108
  temperature, 117
Specific fuel
  consumption -
  see SFC
Spire lock, 132
Splash lubrication,
  129
Spray tube, 86
Sprockets, 142
Square four, 32
Squish, 44
Stoichiometric
  mixture, 83
Straight four, 30
Strain, 169
Strain-hardening,
  169
Stratified charge,
  83
Strength of
  materials, 168
Stress, 169
  concentration,
  173
  raiser, 125, 171
  relief, 129, 171
Stroboscope, 18,
  112
Supercharger, 7
Superflow air

bench, 22
  dynamometer, 24
Surface finish, 174
  plate, 19
Surge, 9
  tank, 99
  valve spring, 71
Suzuki GSX-R750,
  57
  lubrication, 115
  GSX1100E, 8
  XN85, 63
Swirl - see
  Turbulence
Synchronized
  carburettors,
  93
Synthetic oil, 118

Tapered needle, 88
Tappets, 61
Temperature, 174
Testing, 137-142
Tetra-ethyl lead, 13
Thermal efficiency,
  1, 5, 118
Thermostat, 120
Thickness gauge,
  18
Thread lock, 126
Throttle, 88
  response, 90, 140
  stop, 93
  valve, 98
Thrust at rear
  wheel, 150
Tightening torque,
  134
Time-area, 51
Timing, cams, 66
  ignition, 112
  valves, 61
Tools, 16–26
Torque, 2
  wrench settings,
  168
  tightening, 134
Torsional vibration,
  30
Toughness, 170
Track test, 139
Transients, 88, 141
Transmission, 124
Trapping efficiency,
  107
Tri-metal shells,
  127
Triple, 29
Trumpet - see
  bellmouth
Turbocharger, 7

Turbulence, 5, 58
Twin cylinder, 27
  180°, 165
Two-valve head, 48

Ultimate tensile
  strength, 169

V-blocks, 19
V-twin, 28
V-4, 31
V-6, 33
V-8, 33
Vacuum bleed, 97
  fuel tap, 141
  gauge, 18, 92
  -operated tap, 96
  valve, 96
Valve, 51–72
  acceleration, 143
  angle, 45
  clearance, 61,
  136
  duration, 55
  float, 143
  guide, 61, 64
  clearance, 167
  lift, 51, 43, 62
  motion, 71
  overlap, 6
  seat, 52, 63, 65
  spring, 62, 65
  surge, 71
  stem, 63
  time-area, 144
  program, 146
  timing, 53, 61
  types, 53
  velocity, 143
Vegetable oil, 118
Velocity, piston,
  164
  valve, 143
Venturi, 85, 97
Vernier caliper, 17
  sprocket, 67
Vibration, 165
Viscosity, 118
Volumetric
  efficiency, 3, 73,
  107

Waisted valves, 63
Wastegate, 11
Water brake, 20, 24
  injection, 108
Weak mixture, 82
Welded crankpin,
  128
Wet sump, 114
Wills ring - see

177

Gas-filled ring
Wire gauge, 18
Work-hardening, 169

Wrist pin - see Piston pin

Yamaha 608cc

modifications, 145
EXUP, 74
FZ, 57

FZR750/1000, 53
Genesis, 30
Yield point, 169
Young's modulus, 169